TIDAL INLETS

Hydrodynamics and Morphodynamics

This book describes the latest developments in the hydrodynamics and morphodynamics of tidal inlets, with an emphasis on natural inlets. A review of morphological features and sand transport pathways is presented, followed by an overview of empirical relationships between inlet cross-sectional area, ebb delta volume, flood delta volume and tidal prism. Results of field observations and laboratory experiments are discussed and simple mathematical models are presented that calculate the inlet current and basin tide. The method to evaluate the cross-sectional stability of inlets, proposed by Escoffier, is reviewed, and is expanded, for the first time, to include double inlet systems. This volume is an ideal reference for coastal scientists, engineers and researchers, in the fields of coastal engineering, geomorphology, marine geology and oceanography.

J. VAN DE KREEKE is Emeritus Professor at the Rosenstiel School of Marine and Atmospheric Science, University of Miami, where his research focused on coastal engineering and estuarine and nearshore hydrodynamics. He has published extensively on tidal inlets, and is the editor of *Physics of Shallow Estuaries and Bays* (Springer-Verlag, 1986). In 2004, Professor van de Kreeke received the Bob Dean Coastal Research Award for world-renowned research on tidal inlets.

R.L. BROUWER, while at Delft University of Technology, The Netherlands, wrote both his MSc and PhD theses on the subject of cross-sectional stability of double inlet systems. He continued working on this subject as a postdoctoral fellow and at the same time did pioneering work in deploying drones for coastal and inlet research. He has published several papers on double inlet systems in refereed journals and conference proceedings. Presently, he is employed as a senior researcher at Flanders Hydraulic Research in Antwerp, Belgium.

TIDAL INLETS

Hydrodynamics and Morphodynamics

J. VAN DE KREEKE
University of Miami, USA

and

R.L. BROUWER
Delft University of Technology, The Netherlands

CAMBRIDGE
UNIVERSITY PRESS

University Printing House, Cambridge CB2 8BS, United Kingdom

One Liberty Plaza, 20th Floor, New York, NY 10006, USA

477 Williamstown Road, Port Melbourne, VIC 3207, Australia

4843/24, 2nd Floor, Ansari Road, Daryaganj, Delhi – 110002, India

79 Anson Road, #06–04/06, Singapore 079906

Cambridge University Press is part of the University of Cambridge.

It furthers the University's mission by disseminating knowledge in the pursuit of education, learning, and research at the highest international levels of excellence.

www.cambridge.org
Information on this title: www.cambridge.org/9781107194410
DOI: 10.1017/9781108157889

First published 2017

A catalogue record for this publication is available from the British Library.

ISBN 978-1-107-19441-0 Hardback

Contents

Preface

Historically, interest in tidal inlets originates from their importance for commercial shipping and recreational boating. Unfortunately, when in a natural state, most inlets are less than ideal from a navigational point of view and need improvement. They are unstable, i.e., as a result of tide and waves they have a tendency to migrate and shoal. Initially, to stabilize inlets, common sense and practical experience was used as the sole guide. It was not until the nineteen-twenties that research, using field observations, mathematical analysis and laboratory experiments, led to an improved understanding of the complex physical processes that govern the water motion and morphology of tidal inlets. This knowledge could then be used to arrive at science-based improvements.

This book summarizes and synthesizes the scientific advances in inlet research with emphasis on the period 1978 to present. It is a sequel to the earlier books, *Stability of Coastal Inlets* by Per Bruun and Gerritsen (1960) and *Stability of Tidal Inlets: Theory and Engineering* by Per Bruun et al. (1978). The focus is on natural (no man-made modifications) tidal inlets in a sandy environment. The book is intended for anyone who is interested or has dealings with tidal inlets, including coastal engineers, coastal scientists, students and managers. The two authors made an equal contribution to the contents of this book.

Per Bruun and Frans Gerritsen, through their afore mentioned book, were central in introducing Co van de Kreeke to the field of tidal inlets. Discussions with Per Bruun, Robert Dean, Murrough O'Brien and Ashish Mehta have further stimulated this interest. Ronald Brouwer was introduced to the field of tidal inlets by Co van de Kreeke. They worked closely together during his graduate work on tidal inlets at Delft University of Technology, The Netherlands. The support of Henk Schuttelaars and Pieter Roos during that period is acknowledged.

In preparing the manuscript, a number of chapters have benefited greatly from discussions with colleagues. They include Henk Schuttelaars and Pieter Roos on Chapters 9 and 10, Zheng Wang on Chapter 12 and Erroll Mclean and Jon

Hinwood on Chapter 13. Albert Oost was helpful in explaining the geology and sedimentology of the Wadden Sea.

The authors would like to acknowledge Duncan FitzGerald, Todd Walton, Ashish Mehta, Henk Schuttelaars, Marcel Stive, Pieter Roos, Huib de Swart, Zheng Wang and Judith Bosboom for reviewing earlier versions of different chapters.

Co van de Kreeke did most of his research on tidal inlets during the period 1971–2003, while a professor at the Rosenstiel School of Marine and Atmospheric Science of the University of Miami. During that period he cooperated regularly with the National Institute of Coastal and Marine Management of the Dutch Rijkswaterstaat. In that context the many discussions with Job Dronkers should be mentioned. After graduating, Ronald Brouwer worked on the topic of tidal inlets as a Post-Doctoral Fellow at the Delft University of Technology.

In-kind support of the Rosenstiel School of Marine and Atmospheric Science of the University of Miami, by providing work space, computer support and library services during Co van de Kreeke's tenure as an emeritus professor, is acknowledged. Marcel Stive was instrumental in having Ronald Brouwer do his graduate and post-doctoral work on tidal inlets and providing financial support.

We wrote this book out of curiosity, looking for answers to questions such as: why do tidal inlets wander and shift position; why are maximum velocities close to the same for most inlets; how do inlets interact with other inlets; how do they affect the beaches; and many more. We found the answers to some of the questions but certainly not to all of them. There remains room for future research.

Not knowing all the answers made it difficult to decide when and where to stop. We decided that, after having worked on the book for five years, it was enough and could we answer the often-posed question, "are you still working on that book?" with a resounding "No!"

1

Introduction

In the context of this book, tidal inlets are defined as the relatively short and narrow passages between barrier islands. They are sometimes referred to as passes or cuts. Tidal inlets are a common occurrence as barrier island coasts cover some 10 percent of the world's coasts (Glaeser, 1978). According to Hayes (1979), their presence is limited to coasts where the tidal range is less than 4 m.

The earliest interest in tidal inlets originates from their importance to commercial shipping. The relatively protected back-barrier lagoons were a favorite location for harbors. Later, with the increase in recreational boating, small boat basins and marinas were located in back-barrier lagoons. In addition to these commercial and recreational aspects, tidal inlets are ecologically important. Through the exchange of lagoon and ocean water, they contribute to the increase of water quality in the lagoon. Unfortunately, there is also a downside: tidal inlets interrupt the flow of sand along the coast. They not only interrupt but also capture part of the sand, causing erosion of the downdrift coast. For example, in Florida, with some eighty inlets, much of the beach erosion has been attributed to tidal inlets.

Most natural tidal inlets are less than ideal from a navigational point of view. The many shoals, the strong tidal current and the exposure to ocean waves make entering difficult. In addition, on timescales of years to decades, the morphology shows considerable variation, and maintaining sufficient depth and alignment of the channels requires substantial dredging. To minimize dredging and to improve navigation conditions, many inlets have been modified by adding jetties and breakwaters. As a result, tidal currents, waves and sand transport pathways differ from those at inlets without these structures. Nevertheless, in this book, emphasis is on tidal inlets that have not been modified. The reasoning is that understanding the physical processes governing the behavior of tidal inlets in a natural state is a prerequisite for the proper design of engineering measures. This includes the determination of undesirable side effects such as erosion of the adjacent beaches.

The main morphological features of a tidal inlet are the inlet, the ebb delta on the ocean side and the flood delta on the lagoon side. The morphology of the inlet and deltas is shaped by tide and waves. Depending on their relative importance, tidal inlets have been categorized as tide- or wave-dominant (Davis and Hayes, 1984; Hayes, 1994). Tides tend to keep inlets open. In this respect, the tidal prism – the volume of water entering the inlet on the flood and leaving during the ebb – is an important parameter; the larger the tidal prism, the larger the inlet. In turn, waves tend to close the inlet through the wave-driven longshore and cross-shore sand transport.

Scoured in sand, tidal inlets are dynamic features; inlet and channels move, ebb deltas change shape and volume. In discussing these morphological changes, emphasis is on processes with timescales of days (storm timescale), weeks and decades, as opposed to the geological timescale. This excludes the small-scale sand transport processes to which reference is made to Soulsby (1997) and van Rijn (1993). The morphology of the back-barrier lagoon comes into play only as it affects the water motion in the tidal inlet. For the water motion and morphology of the back-barrier lagoon, reference is made to Dronkers (2005).

Examples of barrier island coasts are the East Coast of the US, the Gulf Coast of the US, the Dutch, German and Danish Wadden Sea coast (Fig. 1.1a), the east coast of Vietnam, the northeast coast of the North Island of New Zealand, the Algarve coast of Portugal (Fig. 1.1b) and the Adriatic coast of Italy (Fig. 1.1c). Many inlets along these coasts have been studied extensively. For the origin and morphology of barrier island coasts reference is made to a series of articles in Davis (1994). Rather than one, most barrier island coasts consist of a chain of islands resulting in multiple inlets connected to the same back-barrier lagoon. In fact, it would be difficult to find a tidal inlet that is not affected by a companion inlet. In case of multiple inlets, each inlet competes for part of the tidal prism. This could lead to some inlets closing while others remain open.

Since the 1960s the main tool in studying tidal inlet processes has shifted from laboratory research, including scale models and flume studies, to mathematical models. A distinction is made between process-based models and empirical models (van de Kreeke, 1996). Following Murray (2003), process-based models are divided in exploratory and simulation models. Exploratory models only include the processes that are essential in reproducing the basic behavior. Process-based simulation models start with basic physics and are designed to reproduce the behavior of a natural system, or a schematization thereof, as accurately as possible. Process-based exploratory models are usually simplified to a level allowing analytical or semi-analytical solutions. Although not adequate for predictive purposes, they can be used in a diagnostic mode to help understand phenomena observed in the field

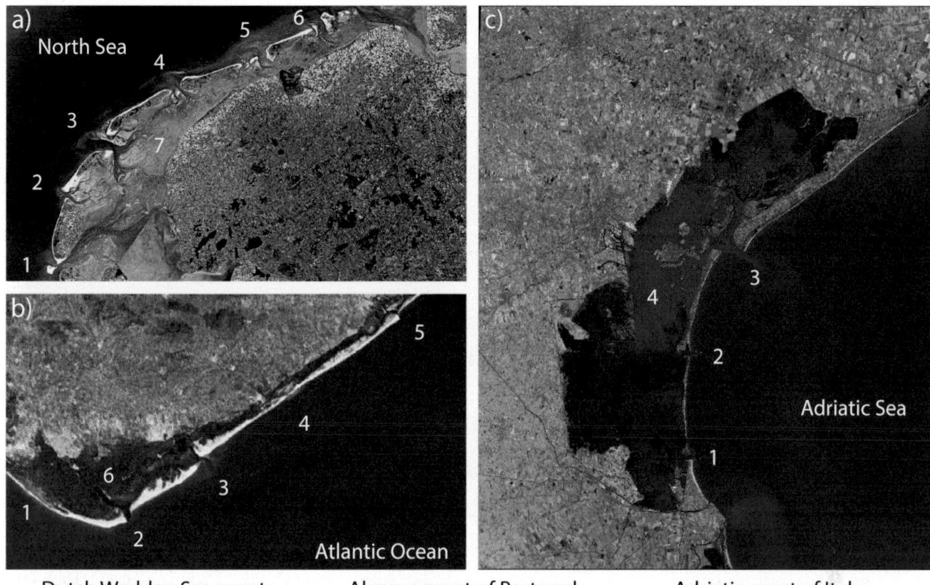

Dutch Wadden Sea coast:
1. Texel Inlet
2. Eyerlandse Gat Inlet
3. Vlie Inlet
4. Ameland Inlet
5. Frisian Inlet
6. Lauwers Inlet
7. Wadden Sea

Algarve coast of Portugal:
1. Ancão Inlet
2. Faro-Olhão Inlet
3. Armona Inlet
4. Fuseta Inlet
5. Tavira Inlet
6. Ría Formosa Lagoon

Adriatic coast of Italy:
1. Chioggia Inlet
2. Malamocco Inlet
3. Lido Inlet
4. Venice Lagoon

Figure 1.1 a) the Dutch Wadden Sea coast (USGS and ESA, 2011), b) the Algarve coast of Portugal (Esri et al., 2016) and c) the Adriatic coast of Italy (NASA et al., 2003).

and to check the validity of the results of the more complicated simulation models. Solving the equations underlying the process-based simulation models requires a numerical approach. The models provide a far more realistic representation of the physical processes than the exploratory models; however, a drawback is that it is often difficult to pinpoint the interactions that determine the overall behavior.

Empirical models are based on the assumption that after a perturbation the morphology tends towards an equilibrium state. The equilibrium state is defined by empirical relationships between the size or volume of the morphological units and the tidal prism. The return to equilibrium is described by empirical equations. Empirical models are a useful substitute when knowledge of the basic processes is insufficient, as is often the case.

The book summarizes and synthesizes the advances in tidal inlet research over the past 40 years. Emphasis is on natural inlets in a sandy environment with tide and waves as the dominant forcing. The book is organized as follows.

In Chapters 2–4 a description of the morphology and morphological changes of tidal inlets is presented. Chapter 2 describes the origin and major elements: inlet, ebb delta, flood delta and back-barrier lagoon. Chapter 3 focuses on sand transport pathways and sand bypassing. Attention is given to location stability, modes of bypassing and their relationship to the ratio of tidal prism and longshore sand transport (P/M ratio). Furthermore, the effect of inlets on the adjacent shores is discussed. In Chapter 4, selected inlets are reviewed with emphasis on sand bypassing and location stability.

Chapter 5 deals with the empirical relationships. The empirical relationship between inlet cross-sectional area and tidal prism (A–P relationship) is discussed and a physical explanation for this relationship is given. Additionally, the concept of equilibrium velocity is introduced. An empirical relationship between delta volume, tidal prism and wave energy is also presented.

Chapters 6 and 7 introduce process-based exploratory models that are used to explore the hydrodynamics of tidal inlets. In Chapter 6 a lumped parameter model and the Keulegan and Öszoy–Mehta Solutions are described. The internal generation of the third harmonic is discussed. Solutions are applied to a representative inlet and results are compared to those of a numerical solution. In Chapter 7 the hydrodynamic equations are expanded to include depth variations with tidal stage. The analytical solution to the expanded equations shows the generation of even overtides and the resulting tidal asymmetry, mean inlet velocity and mean basin level.

Chapters 8–10 deal with cross-sectional stability. Cross-sectional stability is determined using the Escoffier Stability Model. The Escoffier Stability Model, including the Escoffier Diagram, is described in Chapter 8. As examples, the model is applied to two single-inlet systems, Pass Cavallo (TX) and a representative inlet. An expression for the adaptation timescale of the inlet cross-sectional area after a storm is presented. Chapters 9 and 10 deal with the cross-sectional stability of double inlet systems, i.e., rather than one inlet the back-barrier lagoon is connected to the ocean by two inlets. In Chapter 9 the water motion in the inlets is described by the lumped parameter model. This model includes the one-dimensional hydrodynamic equations for the inlet and the assumption of a uniformly fluctuating basin water level. To investigate the effect of this assumption on cross-sectional stability, variations in basin water level are introduced by dividing the basin into two sub-basins connected by an opening, representing a topographic high. As part of the stability analysis a flow diagram is introduced. The flow diagram is the two-dimensional counterpart of the Escoffier Diagram. In Chapter 10, the spatial variations in basin water level are introduced by describing the hydrodynamics of inlets and basin by the shallow water wave equations. A semi-analytical solution is used to solve the governing equations.

Chapters 11 and 12 present applications of a process-based simulation model and an empirical model, respectively. In Chapter 11 a process-based simulation model is used to determine the morphology of a newly opened tidal inlet with emphasis on the inlet and the ebb delta. Using different inlet dimensions, ocean tidal amplitudes and basin surface areas, a series of numerical experiments is carried out to verify the A–P relationship. Chapter 12 describes the use of empirical models to explain the ebb delta development at a newly opened inlet (Ocean City Inlet, MD) and the adaptation of an inlet and ebb delta after basin reduction (Frisian Inlet, NL).

Chapter 13 focuses on the effect of river flow on the entrance stability of tidal inlets. The effect of river flow on the basin tide and the mean basin level is shown by expanding the Öszoy–Mehta Solution to include river flow. The effect of river flow on cross-sectional stability is discussed for a permanently open inlet, a seasonally open inlet and an intermittently open inlet. An exploratory morphodynamic model for the evolution of the depth of an inlet subject to river flow is presented.

Chapter 14 reviews measures to improve navigation and sand bypassing at tidal inlets.

2
Geomorphology

2.1 Introduction

Depending on their origin, tidal inlets are identified as primary or secondary inlets. Regardless of the origin, the morphology is characterized by three major elements: the inlet, the ebb delta and the flood delta. The morphology of each element is determined by tide and waves. In particular, the tidal prism (the volume of water entering on the flood and leaving on the ebb) and the wave-induced longshore sand transport play an important role in determining the cross-sectional area of the inlet and the size and shape of the ebb delta.

2.2 Origin of Tidal Inlets

Following Ehlers (1988), in tracing the origin of tidal inlets a distinction is made between primary and secondary inlets. Primary inlets are those where pre-existing relief, characterized by troughs and adjacent highs, plays a decisive role in the formation of the inlet. During the Holocene, starting some 10,000 BP, onshore sand transport associated with the rapid rise in sea level caused these existing troughs to fill while barrier islands formed on the adjacent highs (Jelgersma, 1983). Examples of primary inlets are the Ameland Inlet (Fig. 4.7) and the Frisian Inlet (Fig. 12.4) along the Dutch Wadden Sea, both relics of drowned river valleys (Beets and van der Spek, 2000).

Apart from these primary tidal inlets, secondary inlets can be identified. Secondary inlets originate from flooding of narrow and shallow parts of barrier islands during a storm. Once the fore-dune ridge is dismantled as a result of storm erosion, a shallow washover channel develops. As the storm passes and winds change direction, return flow forces water against the landward side of the barrier. Often the return flow is funneled across the low portion of the barrier island through the washover channel. Depending on the tidal prism and the longshore sand transport, the washover channel closes or remains open. When remaining open, this channel

Figure 2.1 The breach at Old Inlet, Fire Island (NY) before (left, 2010) and after (right, November 2012) hurricane Sandy (National Park Service, 2012).

is then gradually enlarged by the ensuing tidal currents. Sand from the channel is deposited both offshore and in the basin, forming the onset to, respectively, the ebb delta and the flood delta. All these secondary or washover inlets are located in a sand-rich environment. As discussed in more detail in the following chapters, the ultimate shape and size of newly opened inlets depends on tide and waves.

A description of a washover inlet that ultimately closed on its own can be found in El-Ashry and Wanless (1965). The tidal inlet was located 4.5 miles north of Cape Hatteras (NC). It opened in March 1962 during a severe storm. Large volumes of sand collected at the landward side of the inlet with channels connecting inlet and lagoon. During the first month the inlet width and the volume of the deposits experienced rapid growth, and then the growth slowed down and ultimately the inlet closed on its own. More recently, as a result of hurricane Sandy, a washover inlet was opened at Fire Island (NY), connecting Great South Bay and the ocean (Fig. 2.1). The inlet opened in October 2012 and since then has been extensively monitored. As of September 2016, the inlet is still open (National Park Service, 2012).

In addition to these naturally opened inlets, there are numerous man-made inlets. They are opened or relocated for the purpose of either navigation or water quality. Examples are Government Cut, Lake Worth Inlet, South Lake Worth Inlet and Bakers Haulover Inlet, all located on the southeast coast of Florida (Stauble, 1993). The first two were opened for navigation purposes. Water quality was the main motive for opening South Lake Worth Inlet and Bakers Haulover Inlet.

2.3 Equilibrium Morphology

Newly opened inlets either close or remain open. When remaining open the morphology tends towards equilibrium. In most cases this equilibrium is dynamic rather than static, whereby the morphology oscillates about an equilibrium state.

Oscillations are either the result of variations in the hydrodynamic forcing or are associated with intrinsic instabilities of the morphology.

Examples of variation in forcing are the spring–neap tidal cycle and the seasonal variation in storm severity. Byrne et al. (1974) report fortnightly variations in the cross-sectional area of the inlet channel of Wachapreague Inlet (VA) (Section 4.6), resulting from the spring–neap variation in the tide. Seasonal variations in morphology are reported in Morris et al. (2004). Using video techniques and wave measurements they show that the seasonal behavior of the Barra Nova Inlet (formerly called Ancão Inlet), Rìa Formosa, Portugal, is cyclic in nature. The center and the alignment of the inlet change due to high-energy winter storms, after which the inlet returns to the original morphological state during the remainder of the year. The timing and progression of the morphology through this cycle is thus closely related to the seasonal variation in local wave climate.

Examples of intrinsic instabilities in morphology are spit formation and breaching and the movement of channels on the ebb delta. Both of these lead to cyclic variations in morphology. Examples of tidal inlets where spit formation and breaching plays a role are Captain Sam's Inlet (SC) (Hayes, 1977) and Mason Inlet (NC) (Cleary and FitzGerald, 2003). Channel movement on the ebb delta is observed at Ameland Inlet, The Netherlands (Israel and Dunsbergen, 1999) and Price Inlet (Fitzgerald et al., 1984). In both cases the timescale of the cyclic motion is measured in decades. Detailed information on Captain Sams, Mason, Ameland and Price Inlets is presented in Chapter 4.

In summary, the morphology of tidal inlets is not static but shows variations, with timescales ranging from weeks to decades. In addition, as a result of gradual infilling of the back-barrier lagoon, and depending on the rate of sea level rise, tidal inlets on timescales of centuries can become self-destructive.

2.4 Large-Scale Morphological Elements

In spite of the forcing and sediment conditions being different, the morphology of most tidal inlets is characterized by three large-scale elements: inlet, ebb delta and flood delta (Fig. 2.2). The deltas are an integral part of the tidal inlet as they affect the hydrodynamics and vice versa. In particular, the ebb delta constitutes a bridge that allows bypassing of sand from the updrift to the downdrift coast. In the next sections the main characteristics of each of these elements will be briefly discussed.

2.4.1 Inlet

Here, "inlet" refers to the channel separating two barrier islands. Inlets exhibit a wide range of plan-forms. Some are straight and oriented perpendicular to

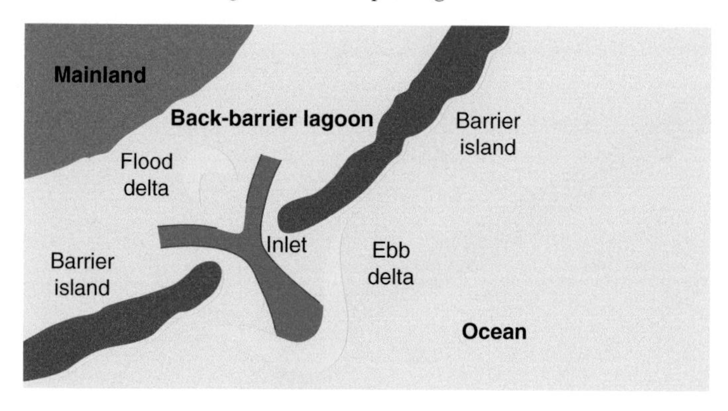

Figure 2.2 Schematized tidal inlet (adapted from Davis and FitzGerald, 2004).

the coast; others exhibit strong curvature. An important role in determining the orientation is played by the longshore sand transport and the orientation of the channels in the back-barrier lagoon. In addition, geological constraints, such as consolidated sediment and rock outcroppings, can play a role. At the seaward end, where the inlet connects to the ebb delta, there is often a tendency for the inlet to branch in separate ebb and flood channels: a result of different flow patterns for ebb and flood. The ebb flow pattern is similar to that of a turbulent jet, whereas the flood flow resembles a streamlined flow into a bell-shaped entrance (Stommel and Farmer, 1952).

Cross-sectional areas of the inlet vary and are assumed to be smallest where the inlet is narrowest and the depth is largest. This part of the inlet is referred to as the gorge or the throat. The cross-sectional area of the throat section is determined by the tidal prism and longshore sand transport (Chapter 5). The shape of the inlet cross-sections is most often asymmetric, being steeper on one side than the other, and the result of the ebb flows having a preference for one of the inlet sides. Dimensions of inlets are: length 500–5,000 m, depth 2–30 m and width 50–2,000 m.

2.4.2 Ebb Delta

An ebb tidal delta, or ebb delta, is the body of sand on the ocean side of a tidal inlet. Typically, ebb deltas consist of a triangular or half-circle-shaped platform: a sandy area of relatively shallow depth containing low-relief bars and shallow channels emanating from the inlet. The shape and volume of the ebb delta are determined by the relative dominance of tidal versus wave energy. In tide-dominated environments, ebb deltas are relatively large, extending far offshore. In wave-dominated environments, ebb deltas are small, hugging the shore (Sha, 1989). In

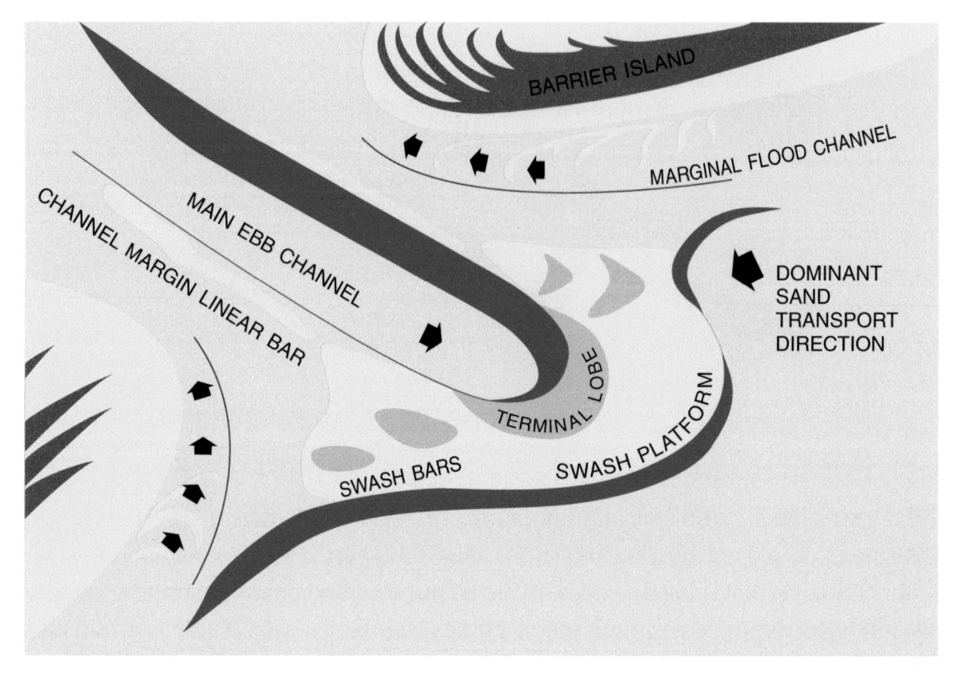

Figure 2.3 Idealized ebb delta (adapted from Hayes, 1980, and Davis and FitzGerald, 2004).

wave-dominated inlets, with a large longshore sand transport, the channels on the platform are forced in a downdrift direction and the delta takes on an asymmetric shape.

Based on field studies in a number of tidal inlets, Hayes (1980) presented a general description of ebb delta morphology. In its simplest form the ebb delta has one main channel conveying the flow from the inlet to the ocean (Fig. 2.3). On both sides of the main ebb channel are channel margin linear bars, levee-like deposits. At the end of the main channel is a relatively steep, seaward-sloping lobe of sand called the terminal lobe. Broad sheets of sand, called swash platforms, flank both sides of the main channel. On the platform are swash bars that form by the swash action of waves. The swash bars migrate across the swash platform by the action of wave-generated currents (King, 1972). The ebb delta is separated from the barrier islands by marginal flood channels. Because this is an idealized representation, some of the features in this simplified model will be more pronounced in some tidal inlets than in others, and some might not be present at all.

The depth of water over the delta affects navigation. In that respect the depth over the shallowest part of the ebb channel is important. It is usually located over the terminal lobe. In general, this depth is much less than the minimum depth in the inlet (Bruun et al., 1978; Dean, 1988).

2.4.3 Flood Delta

Much less is known about flood deltas than ebb deltas. Flood deltas are shaped chiefly by tidal currents as waves play a minor role. At many tidal inlets the flood delta is not even a distinguishable feature as it has merged with the marsh or tidal flats. Flood delta characteristics differ widely from one tidal inlet to another and therefore it is difficult to present their features in a single stylized figure, as was done for the ebb delta. An attempt to identify the major components of flood deltas is described in Hayes (1980). Where flood deltas are a distinct feature, the indications are that their volume increases with increasing values of the tidal prism (Powell et al., 2006).

2.5 Back-Barrier Lagoon

Following Davis and FitzGerald (2004), back-barrier lagoons resemble either an open bay (Fig. 2.4a), an elongated stretch of open water parallel to the mainland (Fig. 2.4b) or a basin with tidal channels marshes and mudflats (Fig. 2.4c). The morphology of the back-barrier lagoon and inlet is shaped by the tide (Dronkers, 2005). In this respect an important role is played by the tidal prism, the volumes of water entering on the flood and leaving on the ebb. In addition to an exchange of water, the tide results in an exchange of sediment. Although the amounts of sediment entering on the flood and leaving on the ebb can be substantial, the net transport over a tidal cycle is usually small, with the direction depending on the tidal velocity asymmetry (Section 7.5). However, on timescales much larger than the tidal period, the net transport can be significant. In case of a net import, sediment is deposited in the relative calm lagoon water resulting in the formation of tidal flats and a decrease in surface area and depth. Where vegetation is established, the originally open lagoon increasingly resembles a marsh area incised with channels. Examples are the tidal inlets along the South and North Carolina coast.

As a result of lagoon infilling, the tidal prism and the velocities decrease. Ultimately, velocities decrease to a level where they are no longer capable of removing

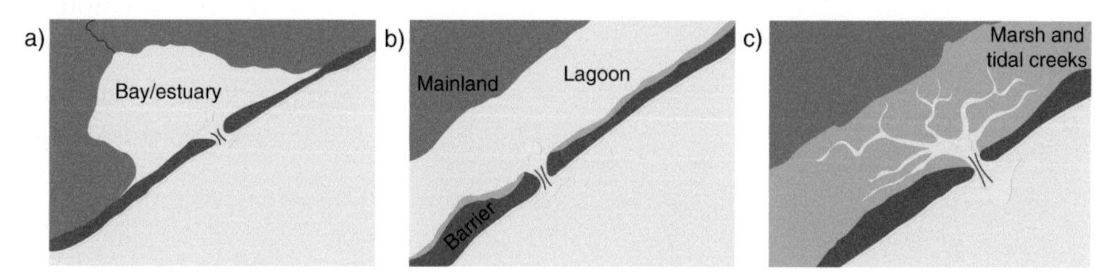

Figure 2.4 Different types of back-barrier lagoons (adapted from Davis and FitzGerald, 2004).

the longshore sand transport and storm deposits out of the inlet, and the inlet closes. Lagoon infilling takes place on timescales of centuries. Therefore, disregarding sea level rise, many inlets on these timescales are self-destructive (van der Spek and Beets, 1992). When accounting for sea level rise, the fate of the tidal inlet depends on the relative rate of sea level rise and the rate of sedimentation in the lagoon. If the sedimentation rate is smaller than the rate of sea level rise, the tidal prism will remain close to the same and the inlet will remain open. When larger than the rate of sea level rise, the lagoon will fill in. As a result, the tidal prism will continue to decrease and the inlet will ultimately close (Louters and Gerritsen, 1994). An example of an ephemeral inlet where the rate of sedimentation was larger than the rate of sea level rise is described in Jelgersma (1983). The Bergen Inlet along the west coast of The Netherlands existed from 5,300 to 3,300 BP and then closed as a result of infilling, in spite of sea level rise.

3

Sand Transport Pathways

3.1 Introduction

The major elements, inlet, ebb delta and flood delta, together with the adjacent coast constitute a sand sharing system (Dean, 1988); sand is transported among these elements by tide- and wave-generated currents. Because at tidal inlets direct measurements are difficult, much of what is known of sediment transport and sediment transport pathways has been inferred from migration and shape of bed forms and swash bars, dredging records, comparison of sequences of bathymetric maps and aerial photographs (Bruun and Gerritsen, 1959; Hanisch, 1981; Hine, 1975).

Sand is transported towards a tidal inlet by longshore currents. Longshore currents and the resulting longshore sand transport result from waves approaching the coast at an oblique angle (Kamphuis, 2006). Some of the longshore sand transport is carried into the inlet by the flood currents and is deposited in the back-barrier lagoon. Another part is jetted to the deeper parts of the ocean and some of it is transported over the ebb delta to the downdrift coast. The sand stored in the lagoon and the deeper parts of the ocean is lost to the littoral zone. As a result, the supply of sand to the downdrift coast is less than the longshore sand transport causing erosion of this part of the coast. The details of the transport of sand from the updrift to the downdrift coasts are discussed in Section 3.3.

An example of sand entering and leaving an inlet is presented in Fig. 3.1. Sand enters through the porous breakwater and is temporarily stored on the updrift side of the inlet in the form of a protruding sand bank. During ebb, sand is carried from the bank in an offshore direction. A similar process was observed in a small inlet in the Bay Islands, Honduras. In that case the clarity of the water and the size of the inlet (width 3 m, depth 0.3 m) made it possible to visually observe the deposition and formation of the sand bank on the updrift side of the inlet and the subsequent removal of some of the sand during ebb.

Figure 3.1 St. Lucie Inlet (FL). Sand is carried to the inlet by the longshore current and deposited on the updrift side of the inlet. Sand is removed from the updrift side of the inlet by the ebb tidal current. Picture taken somewhere in the seventies (Photo: Paul W. Larsen).

In view of the limited knowledge of sand transport and sand transport pathways, the present conceptual models for the sand transport and the sand transport pathways at tidal inlets are somewhat speculative. A basic assumption is that each element (inlet, ebb delta and flood delta), when disturbed, tends towards an equilibrium state. More specifically, when the volume of one element is altered, the whole system responds to restore the original equilibrium by transporting sand among the different elements. This is described in some detail in Chapter 12.

3.2 Sediment Budget

Whatever is known about sediment transport, sediment transport pathways and volume changes can be expressed in terms of a sediment budget (Dean-Rosati, 2005). The sediment budget is a tally of sediment gains and losses, or sources and sinks, within a specified control area over a given time. In general, the seaward boundary of the control area corresponds to the seaward limit of the littoral zone, the zone extending to a depth where sand transport by waves is negligible. This so-called *closure depth* usually is between 10 and 20 m (Kamphuis, 2006; Komar, 1998).

The landward boundary of the control area varies: in some budgets it includes the entire back-barrier lagoon and in others only the flood delta. In the longshore direction the boundary extends to an updrift position, where the shoreline is not measurably affected by the inlet, and similarly for the downdrift position.

The control area is divided in interconnected cells representing such tidal inlet elements as ebb delta, flood delta, inlet and parts of the updrift and downdrift coast. For each cell the sediment balance can be written as

$$Tr_i - Tr_o - \Delta V + P - R = \text{residual}, \tag{3.1}$$

where Tr_i is the sand transport into the cell, Tr_o is the sand transport out of the cell, ΔV is the change in volume, P and R are the amounts of material placed in and removed from the cell and residual is the degree to which the cell is balanced. Provided sufficient information on the sand fluxes at the boundary of the control area with the outside world is available and ΔV, P and R are known, the sand transport between cells can be calculated.

To aid in the construction of a sediment budget and to evaluate the different terms in Eq. (3.1), a computer program, Sediment Budget Analysis System (SBAS), was developed (Kraus and Rosati, 1999; Rosati and Kraus, 1999). SBAS also allows the user to record uncertainty for each value entered in the sediment budget. Then, for each cell, SBAS calculates the root mean square uncertainty. The user can apply the root mean square uncertainty to indicate the relative confidence that can be given to each cell and compare alternatives that represent different assumptions about the sediment budget.

Examples of sediment budgets are those at Grays Harbor (Byrnes et al., 2003) and Faro-Olhão Inlet (Pacheco et al., 2008). Grays Harbor is an estuary on the southwest coast of the state of Washington, USA. The entrance to the estuary has two jetties constructed during the period 1898–1916. A sediment budget covering the period 1987–2002 is presented. The aim of the sediment budget was to document the sediment transport pathways and rates during that period. The control area covers the entrance area, including part of the offshore, the beaches on either side of the inlet. Sediment fluxes at the boundaries with the outside world were partly based on regional transport results and numerical modeling estimates. The control area is divided in nine cells, representing the major morphological elements. For each cell the change in sand volume and the dredging and placement volumes were determined. With the known sediment fluxes at the boundaries with the outside world, the sand transport between cells follows from the observed volume changes and the known dredging and placement volumes.

Faro-Olhão is one of the inlets of the Ría Formosa (Portugal) barrier island system, serving the cities of Faro and Olhão. The inlet was opened in 1929 and gradually improved with jetties between 1929 and 1955. Sediment budgets were constructed for the time periods 1962–1978 and 1978–2001. The objective was to explain the observed coastline changes after the opening and the stabilization of the inlet. The control area covered the entire inlet system. The seaward limit of the control area was taken at the transition of the ebb delta slope and sea floor. The landward limit of the control area coincides with the seaward boundary of the salt-marsh in the back-barrier lagoon. The control area was divided into six cells representing the flood delta, ebb delta, inlet channel, ebb delta channel and adjacent updrift and downdrift coast. For each cell the change in volume, including the dredged and placed sediment volumes, was determined. Because the information on sediment fluxes at the boundaries of the control area is insufficient, internal sand transport rates and pathways could not be calculated. In the paper considerable attention is given to the different methods to estimate the volume changes in a cell, including the application of the error analysis presented in (Dean-Rosati, 2005).

3.3 Sand Bypassing

3.3.1 Bypassing Modes

The way sand is transported from the updrift to the downdrift coast of a tidal inlet is commonly referred to as inlet sand bypassing. It is the way that sand, after a short interruption on the ebb delta and in the inlet, is returned to the littoral zone. Bruun and Gerritsen (1959) were the first to address bypassing. They reasoned that the transfer of sand is the result of waves and tidal currents. Based on this observation they distinguished between bar bypassing, where waves are dominant, and tidal flow bypassing, where tide is dominant.

Bar bypassing implies that sand is directly transferred from the updrift coast onto the ebb delta and then to the downdrift coast via the terminal lobe and swash platform (Fig. 2.3). The terminal lobe and swash platform serve as a bridge over which the sand is carried to the downdrift beach. When channels are present on the delta platform, they are small and shallow.

Tidal flow bypassing implies that, during flood, sand is transported into the inlet and main ebb channel. Part of this transport takes place across the channel margin linear bars (Fig. 2.3).This lateral inflow of sand often results in a cyclic variation of the orientation of the ebb channel on a timescale of decades. Sand is transported out of the main ebb channel by the ebb tidal currents and deposited at the distal end (the seaward end) of the channel. Swash bars are formed at the distal end that, as a result of waves and the dominant landward current, move onshore over the swash platform. On their way to the beach they form bar complexes. It takes these bar complexes, any time from several years to decades to reach the beach.

Lately, it has become apparent that there are many bypassing mechanisms that do not fit the mode of bar bypassing or tidal flow bypassing. Some eight of these are categorized in FitzGerald et al. (2000). In addition to bar bypassing and tidal flow bypassing, the ones that stand out are spit formation and ebb delta breaching.

Spit formation implies that the littoral drift, rather than being transferred to the ebb delta, is deposited as a sand-spit in front of the inlet. When the spit grows, it forces the inlet and the ebb delta to move in a downdrift direction. When breached, usually at the original position of the inlet, the spit welds onto the downdrift shore together with the sand from the abandoned ebb delta. At many inlets this process is repeated at decadal timescales, resulting in an episodic transfer of large volumes of sand to the downdrift beaches. In Friedrichs et al. (1993), based on a study at Chatham Inlet (MA), it is suggested that spit breaching is initiated by a washover forced by a storm. For the washover to evolve into a channel requires a certain head difference between the inlet and the ocean at the location of the washover. This head difference between inlet and ocean results from the distortion of the tidal wave in the inlet channel and increases with increasing spit length.

Ebb delta breaching involves transfer of sand over the ebb delta. Sand enters the updrift side of the main ebb channel, forcing it to migrate in a downdrift direction. When the inlet throat position is relatively fixed, this causes an increased curvature in the alignment of the channel. A new channel is formed and the shoal between the old and new channel is transported by waves towards the downdrift coast. At many inlets this process is repeated at annual timescales, resulting in an episodic transfer of large volumes of sand to the downdrift beaches.

Because each inlet has its own peculiarities, it is difficult to arrive at a universal framework for bypassing modes. However, the four bypassing mechanisms, bar bypassing, spit formation and breaching, ebb delta breaching and tidal flow bypassing, are believed to at least capture the gross characteristics of inlet bypassing. Examples of inlets with different bypassing modes are presented in Chapter 4.

3.3.2 Bypassing Modes and the P/M *Ratio*

Starting with Bruun and Gerritsen (1959), attempts have been made to correlate bypassing mechanisms with the P/M ratio, where P is the tidal prism under spring tide conditions (m^3) and M is the gross longshore sand transport (m^3 year^{-1}). For a large number of both improved and unimproved inlets Bruun et al. (1978) determined the sand bypassing mode (bar bypassing or tidal flow bypassing) together with values of the ratio of tidal prism and gross longshore sand transport; see columns 2 and 3 in Table 3.1. Based on these observations they concluded that for bar bypassing the P/M ratio had to be smaller than approximately 50 and for tidal flow bypassing the P/M ratio had to be larger than approximately 150.

Recently, for a number of inlets more detailed information on bypassing mechanisms and P/M ratio has become available. The results are summarized in

Table 3.1 *Sand bypassing mode,* P/M *ratio and location stability for selected inlets (Bruun et al., 1978).*

Inlet	Sand bypassing mode	P/M	Location stability
Aveiro Inlet (Portugal)	Not known	60	Fair/Poor
Big Pass (FL)	Not known	100	Fair
Brielse Maas (Neth.)	Bar bypassing	30	Poor
Eyerlandse Gat (Neth.)	Tidal flow bypassing	200	Good
Figueira da Foz (Portugal)	Bar bypassing	28	Poor
Texel Inlet (Neth.)	Tidal flow bypassing	1,000	Good
Vlie Inlet (Neth.)	Tidal flow bypassing	1,000	Good
John's Pass (FL)	Tidal flow bypassing	140	Fair
Longboat Pass (FL)	Tidal flow bypassing	200	Good
Oregon Inlet (NC)	Not known	60	Fair/Poor
Ponce de Leon Inlet (FL)	Bar bypassing	30	Poor

P is spring tidal prism (m^3 $year^{-1}$) and M is gross longshore sand transport (m^3 $year^{-1}$). Values are of overall character.

Table 3.2 *Sand bypassing mode,* P/M *ratio and location stability for inlets discussed in Chapter 4.*

Inlet	Sand bypassing mode	P/M	Location stability
Captain Sam's Inlet (SC)	Spit formation	12	Poor
Mason Inlet (NC)	Spit formation	17	Poor
Breach Inlet (SC)	Ebb delta breaching	72	Fair
Price Inlet (SC)	Tidal flow bypassing	80	Good
Wachapreague Inlet (VA)	Tidal flow bypassing	182–455	Good
Ameland Inlet (Neth.)	Tidal flow bypassing	960	Good
Katikati Inlet (New Zealand)	Tidal flow bypassing	>190	Good

P is spring tidal prism (m^3 $year^{-1}$) and M is gross longshore sand transport (m^3 $year^{-1}$). Values are of overall character.

Table 3.2. For *tidal flow bypassing* the P/M ratios conform to those proposed by Bruun et al. (1978) and are larger than 150. An exception is Price Inlet; the reason for this could be that this inlet is eroded into semi-consolidated Pleistocene deposits (FitzGerald, 1984). The bypassing mode *spit formation and breaching* has the smallest P/M values and the P/M value for *ebb delta breaching* is somewhere in between. No information on inlets with *bar bypassing* is available other than that provided in Bruun et al. (1978). To determine whether bypassing mechanisms are uniquely determined by the P/M ratio, and, if not, what other parameters play a role, additional observations are needed.

3.4 Inlet Closure

Not all sand is necessarily bypassed and some of it is permanently deposited in the inlet, resulting in closure. Following Ranasinghe and Pattiaratchi (2003), a distinction is made between closure by longshore and cross-shore sand transport. Longshore sand transport results from waves approaching the shore at an oblique angle. Onshore transport is due to swell waves approaching the shore more or less perpendicularly. As shown in Section 3.3, inlets with a relatively large longshore transport rate and small tidal prism, and thus a small P / M ratio, are prone to spit formation. If not breached, the spit will continue to accrete and prograde, resulting in a lengthening of the inlet in a shore-parallel direction. The inlet becomes less hydraulically efficient and closes. An example is Midnight Pass on the west coast of Florida (Davis et al., 1987). Assuming the longshore sand transport is small, persistent onshore sand transport by swell waves can become the dominant closure mechanism. An example is Wilson Inlet on the southwest coast of Australia. This inlet is usually closed for a period of 6–7 months every year due to the formation of a sandbar across the inlet (Ranasinghe and Pattiaratchi, 1999). Additional information on Wilson Inlet can be found in Chapter 13.

3.5 Location Stability

As part of the bypassing process, many inlets and channels on the ebb delta migrate. This migration is undesirable from a navigation point of view and has led to the concept of location stability. Location stability refers to the permanence of the location of the inlet and ebb delta channels. Depending on the degree of location stability, Bruun et al. (1978) used the somewhat subjective designations: poor, fair and good. Using these designations, the location stability for the inlets considered in Bruun et al. (1978) and those discussed in Chapter 4, are presented in the last columns of, respectively, Tables 3.1 and 3.2.

From the discussion on sand bypassing in Section 3.3.1, it is evident that location stability is closely related to the way sand bypasses the inlet. In case of *bar bypassing*, there are few navigable channels and, where present, they tend to shift. The location stability is poor. The bypassing mode *spit formation and breaching* involves channel migration but to a lesser extent than for bar bypassing, and the designation poor to fair seems appropriate. When bypassing is by *ebb delta breaching* the throat position is stable but the channels on the ebb delta migrate on a relatively short timescale (years); the designation fair seems applicable. In case of *tidal flow bypassing* the inlet throat position is stable and the channels on the delta show only a slow decadal change in orientation, corresponding to good location stability. The relationship between location stability, mode of bypassing and P/M ratio is reflected in the information presented in Tables 3.1 and 3.2.

3.6 Effect of Inlets on Adjacent Shoreline

From the foregoing, it is obvious that tidal inlets have potential impact on the adjacent shoreline. The nature and severity of the impact, among other factors, depends on the sand bypassing mode. Bar and tidal flow bypassing are more or less continuous processes, whereas spit-formation and ebb delta breaching are discontinuous or intermittent processes. For each of these a simple mathematical model is presented that is helpful in interpreting observed shoreline changes.

3.6.1 Continuous Bypassing

Here, shoreline changes are measured with respect to a x–y coordinate system (Fig. 3.2). The x-axis is in the general direction of the shoreline and $x = 0$ is at the inlet axis. The y-axis is positive in the seaward direction. In case of continuous bypassing the shoreline change is usually in the form of advancement and recession of, respectively, the updrift and downdrift shorelines; or, in unusual cases, both sides can advance or retreat. For a given time, the shoreline change signature includes an even (symmetric) and odd (anti-symmetric) component with the inlet

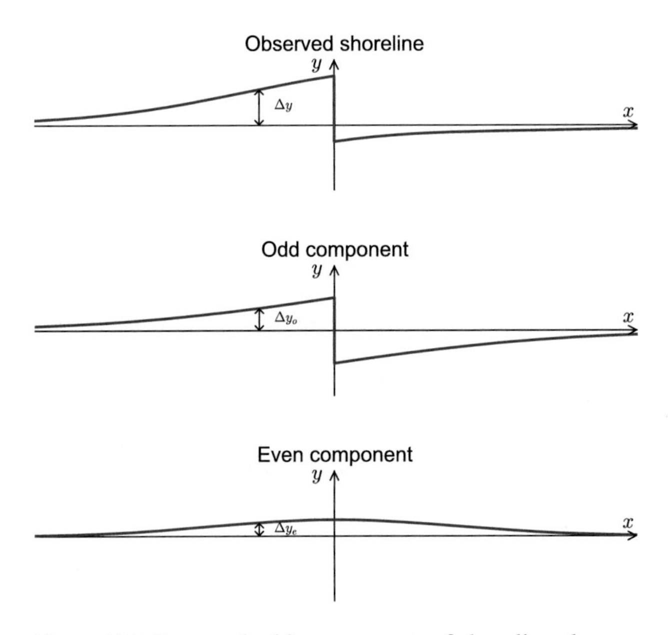

Figure 3.2 Even and odd components of shoreline change.

channel axis as the symmetry axis. The odd component is attributed to the interruption of the longshore sand transport while the even component is attributed to cross-shore transport.

Given the total shoreline change, Dean and Work (1993) describe a procedure, referred to as *even–odd analysis*, to separate the even and odd components (Fig. 3.2). The even–odd analysis is demonstrated, assuming a known total shoreline change $\Delta y(x)$. The total shoreline change is composed of an even component, $\Delta y_e(x)$, and an odd component, $\Delta y_o(x)$, i.e.,

$$\Delta y(x) = \Delta y_e(x) + \Delta y_o(x). \tag{3.2}$$

By definition

$$\Delta y_e(x) = \Delta y_e(-x), \qquad \text{and} \tag{3.3}$$

$$\Delta y_o(x) = -\Delta y_o(-x). \tag{3.4}$$

In terms of the total shoreline change, the even component is

$$\Delta y_e(x) = \frac{\Delta y(x) + \Delta y(-x)}{2}. \tag{3.5}$$

The odd component in terms of the total shoreline change is

$$\Delta y_o(x) = \frac{\Delta y(x) - \Delta y(-x)}{2}. \tag{3.6}$$

Even–odd analysis can be a useful procedure to quantify and interpret the impacts of tidal inlets on the adjacent shores. Carrying out the even–odd analysis for different time periods in combination with information on waves could potentially provide information on the processes that are responsible for the shoreline changes. For a number of improved tidal inlets along the lower east coast of Florida, results of the even–odd analysis are presented in Dean and Work (1993). Although these examples pertain to improved tidal inlets, the method should equally well apply to unimproved tidal inlets. For the jettied tidal inlet at Cape Canaveral Inlet (FL) the results of the even–odd analysis are presented in Fig. 3.3.

3.6.2 Intermittent Bypassing

An example of intermittent bypassing is the Vlie Inlet in The Netherlands (Bakker, 1968). Similar to the Ameland Inlet described in Section 4.8, waves and flood currents transport sand from the updrift coast into the channel across the ebb delta. Ebb tidal currents then are responsible for the transport of the sand to the distal end of the channel, where it forms a shoal. From here the sand is transported by waves onto the ebb delta platform, where it forms large bar complexes. The onshore migration of the bar complexes is intermittent with a timescale of 50–60 years.

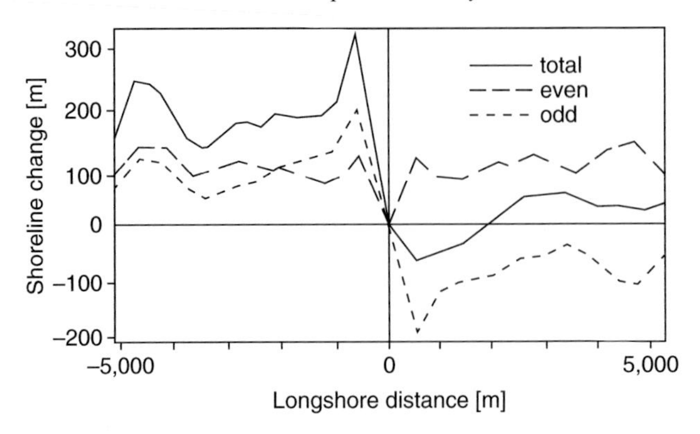

Figure 3.3 Even–odd analysis for Cape Canaveral Inlet (FL) (from Dean and Work, 1993).

The intermittent supply of sand is transported along the downdrift coast in the form of a damped progressive wave. Assuming the transport of sand is by waves, the shoreline position $y(x,t)$ satisfies the diffusion equation (Bakker, 1968, 2013)

$$\frac{\partial y}{\partial t} = \frac{q}{D}\frac{\partial^2 y}{\partial x^2}, \tag{3.7}$$

with y positive in the seaward direction. The direction of the x-axis coincides with the general direction of the coastline and is positive in the downdrift direction. The origin of the x-axis is at the downdrift side of the inlet. The parameter q is a constant depending on wave height, wave period and sand grain characteristics and D is the sum of the closure depth and the berm height. The closure depth is the depth at the seaward limit of the longshore sand transport.

The damped progressive wave, propagating in the positive x-direction, satisfying Eq. (3.7) is

$$y = Ae^{-kx}\cos(\omega t - kx), \tag{3.8}$$

where ω is the radian frequency of the sand wave, k is the wave number and A is the amplitude of the sand wave to be determined from the boundary conditions. The wave number is related to the radian frequency by

$$k = \sqrt{\frac{\omega D}{2q}}. \tag{3.9}$$

At the Vlie inlet, the radian frequency corresponding with the intermittent sand supply is $\omega = 0.11$ rad/year. Observations of coastline positions 4,000 m apart show attenuation in coastline position of 75 percent. Using Eq. 3.8, this translates into $e^{-k(x_2-x_1)} = 0.25$. With $(x_2 - x_1) = 4,000$ m, it follows that $k = 3.6 \times 10^{-4}$ m^{-1}. The corresponding wave length is 1,745 m. The celerity of the sand wave

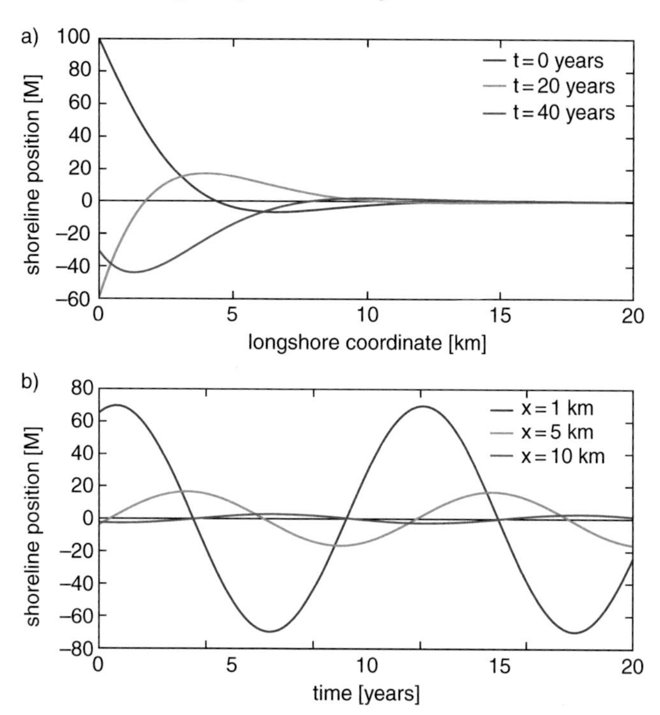

Figure 3.4 Propagating sand wave. a) Shoreline position as a function of longshore coordinate for $t = 0$ years, $t = 20$ years and $t = 40$ years and b) shoreline position as a function of time at $x = 1$ km, $x = 5$ km and $x = 10$ km. Amplitude at $x = 0$ is $A = 100$ m.

$\omega/k = 310$ m/year. Based on observations extending over a period of 70 years, the amplitude of the sand wave at the downdrift side of the inlet ($x = 0$) is $\mathcal{O}(100$ m). As an example, for different years the shoreline positions are presented in Fig. 3.4a and for different longshore positions are presented in Fig. 3.4b.

4

Sand Transport and Sand Bypassing at Selected Inlets

4.1 Introduction

This chapter describes sand transport patterns and sand bypassing at seven inlets; five of these are located on the east coast of the USA (Price Inlet, Breach Inlet, Captain Sam's Inlet, Mason Inlet and Wachapreague Inlet), one inlet is located in the Bay of Plenty on the North Island of New Zealand (Katikati Inlet) and another is part of the Dutch Wadden Sea coast (Ameland Inlet). The inlets are selected because they are still in their natural state and have been extensively studied. Emphasis is on the mode of bypassing, location stability and their relationship with the P/M ratio. In judging the results, it should be pointed out that estimates of longshore sand transport have limited accuracy.

4.2 Price Inlet

Price Inlet (Fig. 4.1) is located on the coast of South Carolina. Tides are semi-diurnal with a mean tidal range of 1.5 m and a spring tidal range of 2.1 m. The annual average deep water significant wave height is 0.6 m. The mean tidal prism is 14×10^6 m^3 and the spring tidal prism is 20×10^6 m^3. The throat cross-sectional area is 894 m^2. From this, maximum cross-sectionally averaged velocities are 1.1 m s^{-1} for mean tide conditions and 1.56 m s^{-1} for spring tide conditions.

The ebb delta has a volume of 6×10^6 m^3 and extents approximately 800 m offshore. The gross longshore sand transport is 0.25×10^6 m^3 year^{-1} and is predominantly from the north. Tide- and wave-generated currents carry the sand through marginal flood channels and across the channel margin linear bars to the main ebb channel. The lateral inflow of sand causes the channel to meander on timescales of decades. With ebb currents stronger than flood currents, most sand deposited in the ebb channel is ultimately transported to the seaward portion of the ebb delta. At low tide, waves break on the seaward edge of the delta and transport sand along the periphery of the delta towards the downdrift beaches and onto the

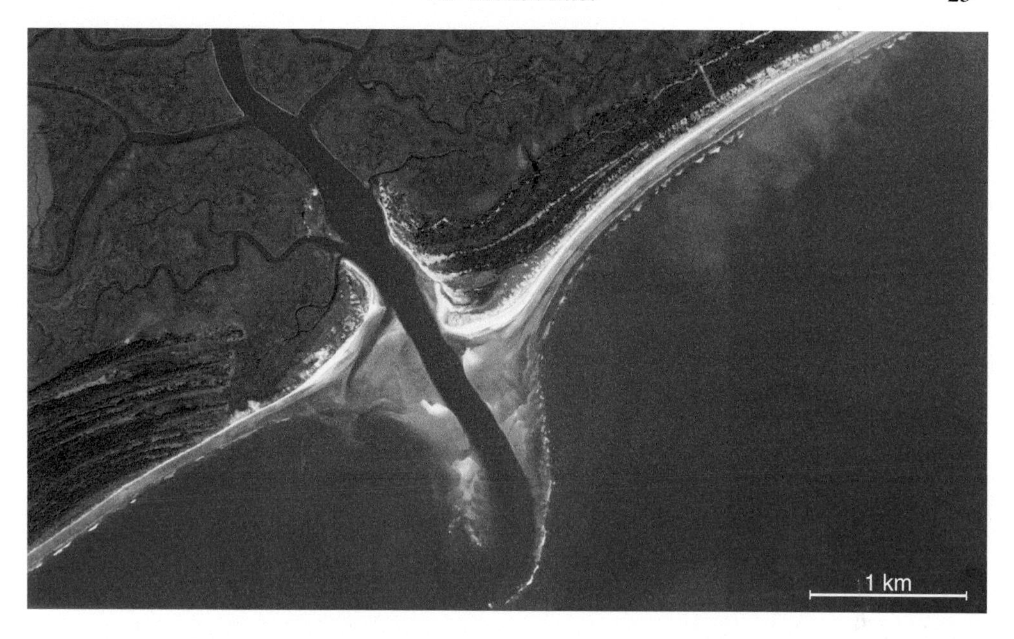

Figure 4.1 Price Inlet (SC) in 2004 (Source: Google Earth).

ebb delta platform. The sand on the ebb delta platform is transported in the form of swash bars. Swash bars travel faster in deeper water than in shallower water and as a result coalesce, forming bar complexes that migrate and attach to the beach. The inlet channel has not migrated, partly because the channel (8.5 m deep) is eroded into semi-consolidated Pleistocene deposits.

Based on these observations it is concluded that the sand bypassing mode is tidal flow bypassing and location stability is good. With a spring tidal prism of 20×10^6 m^3 and a gross longshore sand transport of 0.25×10^6 m^3 year^{-1}, the ratio $P/M = 80$.

Information on Price Inlet is based on Fitzgerald et al. (1984) and Gaudiano and Kana (2001).

4.3 Breach Inlet

Breach Inlet (Fig. 4.2) is located on the South Carolina coast, 19 km south of Price Inlet. The mean tidal range is 1.5 m and the spring tidal range is 2.1 m. The annual average deep water significant wave height is 0.6 m. The tidal prism for mean tide conditions is 13×10^6 m^3 and for spring tide conditions is 18×10^6 m^3. The throat cross-sectional area is 946 m^2. Based on this, maximum cross-sectionally averaged velocities are 0.96 m s^{-1} for mean tide conditions and 1.33 m s^{-1} for spring tide conditions.

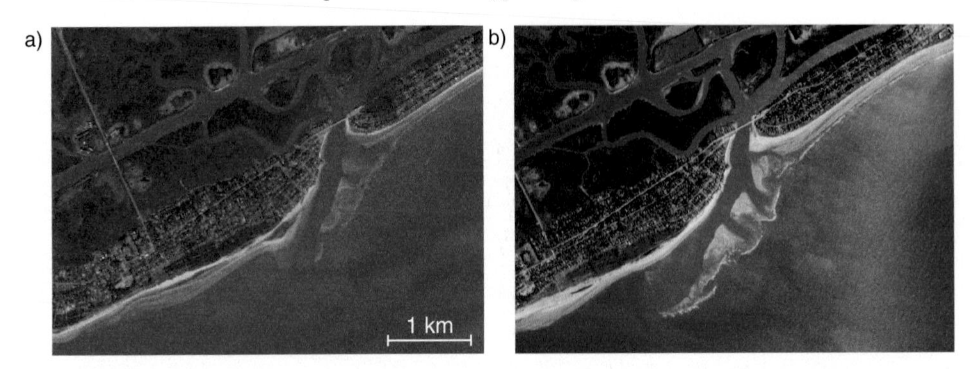

Figure 4.2 Breach Inlet (SC) in a) 2013 (Source: Google Earth) and b) 2015 (Esri et al., 2016).

The ebb delta volume is 7×10^6 m^3. The gross longshore transport is 0.25×10^6 m^3 year^{-1} and is predominantly from the north. The position of the inlet channel has been relatively stable. This is partly attributed to stabilizing structures along the downdrift margin of the inlet. The sequence of bypassing starts with a single channel on the delta (Fig. 4.2a). As a result of the longshore transport the channel and channel-margin linear bar are gradually pushed in a downdrift direction increasing the channel curvature. Because of the increased curvature, the channel becomes hydraulically less efficient and the flow is diverted across the channel-margin linear bar, scouring a new channel (Fig. 4.2b). As a result of wave-generated currents the shoal on the downdrift side of the new channel migrates in a downdrift direction, thereby filling the abandoned channel in the lee of the shoal. The shoal then gradually migrates in a shoreward direction by waves and ultimately welds onto the beach. Shoal attachment is intermittent with an average frequency of once every five years.

Based on the aforementioned, it is concluded that the sand bypassing mode is ebb delta breaching and location stability is fair. With a spring tidal prism of 18×10^6 m^3 and a gross longshore sand transport of 0.25×10^6 m^3 year^{-1}, the ratio $P/M = 72$.

Information on Breach Inlet is based on Fitzgerald et al. (1984) and Gaudiano and Kana (2001).

4.4 Captain Sam's Inlet

Captain Sam's Inlet (Fig. 4.3), also known as Kiawah River Inlet, is located on the central South Carolina coast, approximately 38 km south of Breach Inlet. Tides are semi-diurnal with a mean tidal range of 1.5 m and a spring tidal range of 2.1 m. The annual average deep water significant wave height is 0.6 m. The tidal prism for

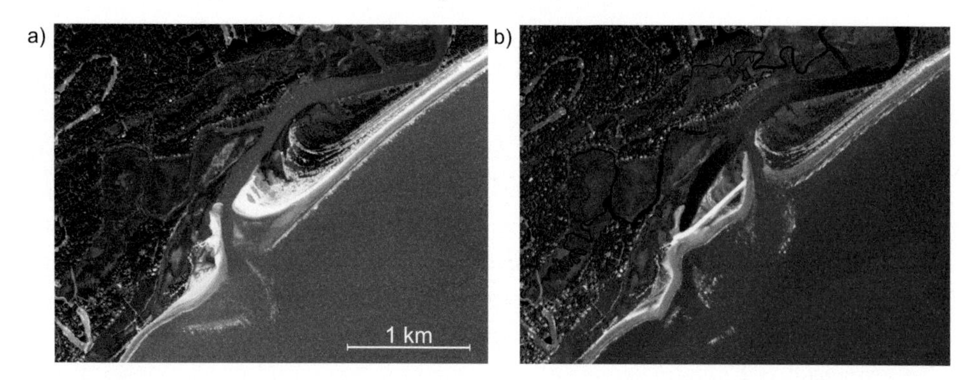

Figure 4.3 Captain Sam's Inlet (SC) in a) 2014 and b) 2015 (Source: Google Earth).

mean tide conditions is 2.3×10^6 m^3 and for spring tide conditions is 3.0×10^6 m^3. The throat cross-sectional area is 210 m^2. From this, it follows that the maximum velocity for mean tide conditions is 0.77 m s^{-1} and for spring tide conditions is 1.0 m s^{-1}.

The ebb delta volume is 4.0×10^6 m^3. The gross longshore sand transport is 0.25×10^6 m^3 year^{-1}. Captain Sam's Inlet has a history of spit formation, elongating the inlet channel and forcing the entrance together with the ebb delta to move in a southerly direction. The average growth rate of the spit is approximately 75 m year^{-1}. As a result of washovers, the spit is periodically breached at the neck. After breaching, the inlet resumes its southerly migration. The abandoned spit and ebb delta move in a southwesterly direction and ultimately weld on to the downdrift beach.

The most recent natural breach at Captain Sams occurred in 1948. The trapping of sand in the spit after this breach caused severe erosion of the downdrift beaches. To alleviate the beach erosion, in 1983 the spit was artificially breached and the abandoned channel was closed. By that time the spit was approximately 1,500 m long and had an average width of 250 m. The shortened inlet was left to adjust naturally by currents and waves. In a period of 3–4 months the initial throat cross-sectional area of 112 m^2 increased to a relatively stable value of 210 m^2. The major impact of the artificial breaching was onshore migration of the abandoned spit and ebb delta. By 1987 most of this sand, estimated to have a volume of 2.0×10^6 m^3 comprising the spit and ebb delta, welded onto the downdrift beaches. After the artificial breaching in 1983, spit extension continued until 1996 when it was artificially breached again. By that time the inlet had moved 1,000 m in a southerly direction.

Based on the foregoing account and inspection of a series of historical photographs, covering the period 1993–2013 (Fig. 4.3), it is concluded that the sand bypassing mode is spit formation and breaching. Smaller volumes of sand are bypassed during breaching events of the main channel of the ebb delta. Location stability is considered poor. With a spring tidal prism of 3.0×10^6 m^3 and a gross longshore sand transport of 0.25×10^6 year^{-1}, the ratio $P/M = 12$.

The artificial breaching of the spit in 1983 and 1996 offered the opportunity to study spit formation and breaching as a mode of sand bypassing in more detail. Morphological changes were documented and from this sediment pathways were inferred. In addition, the experiment provided information on adaptation timescales of inlet cross-section and ebb delta growth.

Information on Captain Sam's Inlet is based on Fitzgerald et al. (1984), Kana and Mason (1988), Kana (1989) and Kana and McKee (2003).

4.5 Mason Inlet

Mason Inlet (Fig. 4.4) is located on the North Carolina coast. Along this part of the coast tides are semi-diurnal with a mean tidal range of 1.0 m and a spring tidal range of 1.2 m. The annual average deep water significant wave height is 0.8 m.

The inlet has a history of migrating to the south, and by 2001 it threatened a resort complex. This condition and the overall degradation of the inlet, including the navigable channels, led to relocation of the inlet and dredging of the back-barrier channels in March 2002. Relocation and dredging resulted in an increase in

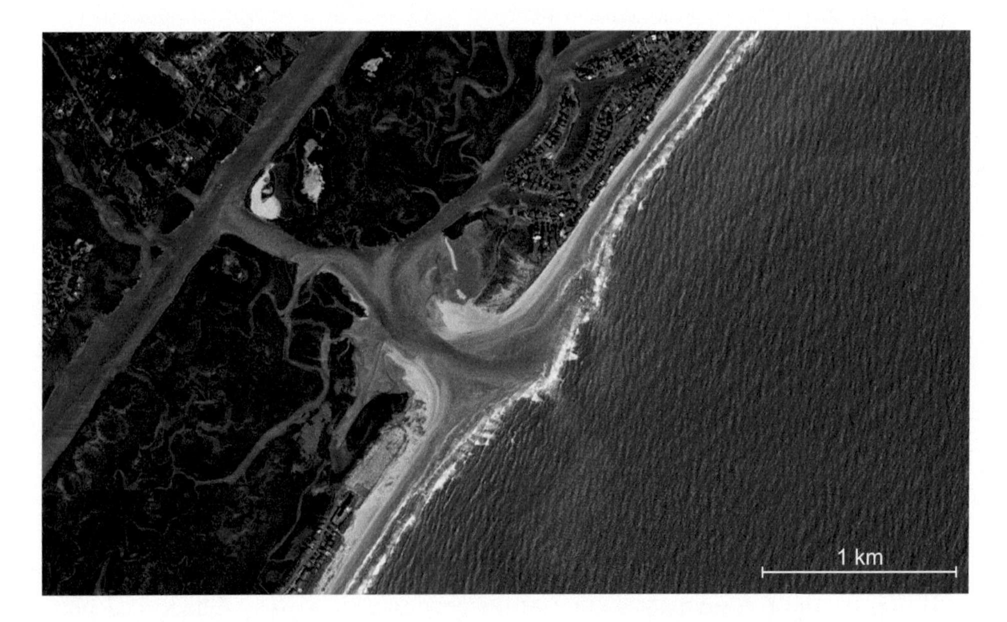

Figure 4.4 Mason Inlet (NC) in 2015 (Source: Google Earth).

mean tidal prism from 0.65×10^6 m^3 prior to 4.2×10^6 m^3 after relocation. The spring tidal prism after relocation is estimated at 5.0×10^6 m^3. The throat cross-sectional area is approximately 300 m^2. From this, it follows that the maximum cross-sectionally averaged velocity for mean tide conditions is 0.97 m s^{-1} and for spring tide conditions is 1.16 m s^{-1}.

The measured ebb delta volume is 0.29×10^6 m^3. The net longshore sand transport is to the south at a rate of approximately 0.3×10^6 m^3 year^{-1}. Because of the persistent southerly direction, the gross longshore sand transport is estimated to be close to the value of the net longshore sand transport.

Based on the foregoing and inspection of a sequence of photographs, covering the period 2002–2013 (Google Earth), it is concluded that the bypassing mode is spit formation and breaching and location stability is poor. With a spring tidal prism of 5.0×10^6 m^3 and a gross longshore sand transport of 0.3×10^6 m^3/year, the ratio $P/M = 17$.

Information on Mason Inlet is based on Cleary and FitzGerald (2003) and Welsh and Cleary (2007).

4.6 Wachapreague Inlet

Wachapreague Inlet (Fig. 4.5) is one of the many inlets along the Atlantic coast of the Delmarva Peninsula (VA). Tides are semi-diurnal with a mean tidal range of 1.16 m and a spring tidal range of 1.37 m. The annual average deep water significant wave height is estimated to be close to the value at Mason Inlet, i.e., 0.8 m. The tidal prism corresponding to the mean tidal range is 77×10^6 m^3 and the tidal prism corresponding to the spring tide is 91×10^6 m^3. The throat cross-sectional area is 4,400 m^2. From this, the maximum cross-sectionally averaged velocity for mean tide condition is 1.22 m s^{-1} and for spring tide condition is 1.44 m s^{-1}.

Wachapreague Inlet connects the coastal ocean to a back-barrier lagoon, consisting largely of tidal flats and marshes. The inlet has a pronounced ebb delta with a single channel. The position of the inlet and the channel has been stable, which in part can be attributed to the presence of cohesive lagoonal mud on the south side of the inlet. Estimates of longshore sand transport, which is primarily from the north, vary between 2×10^5 and 5×10^5 m^3/year (Byrne, personal communication). Based on sequential bathymetric surveys, observed flow patterns and wave information, a qualitative model of sand transport pathways was developed. Longshore sand transport enters the channel from the north. Some of the sand enters the back-barrier lagoon and some of it is transported to the distal end of the channel on the ebb delta. Part of the sand transported to the distal end continues its path along the coast and another part is caught in a sediment loop and returns to the channel.

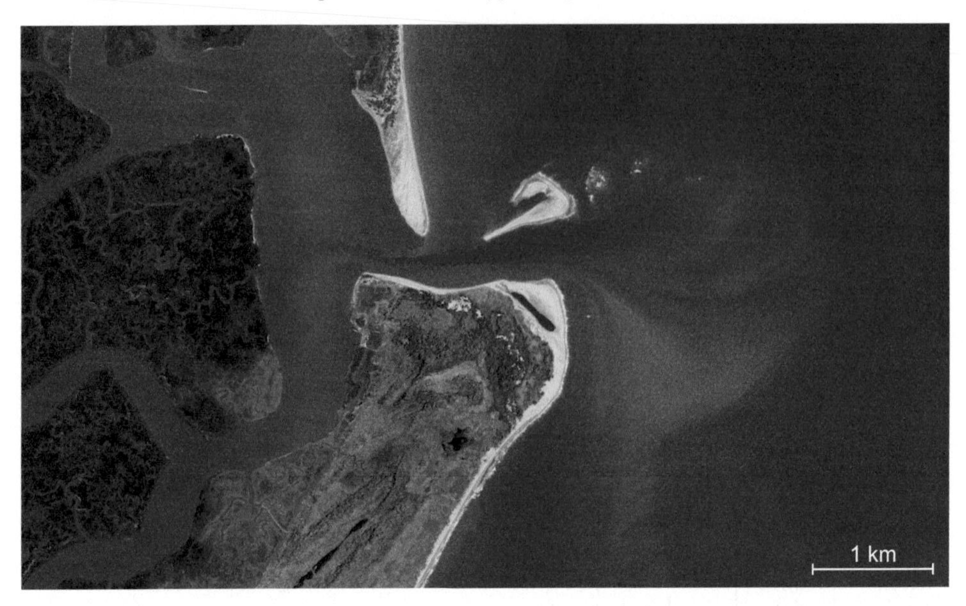

Figure 4.5 Wachapreague Inlet (VA) in 2006 (Source: Google Earth).

Based on these observations, the sand bypassing is by tidal flow bypassing and location stability is good. With a spring tidal prism of 91×10^6 m^3 and depending on the adopted value of the longshore sand transport, the P / M ratio varies between 182 and 455.

Information on Wachapreague Inlet is based on Byrne et al. (1974, 1975), and DeAlteris and Byrne (1975).

4.7 Katikati Inlet

Katikati Inlet (Fig. 4.6) is located in the Bay of Plenty on the North Island of New Zealand. Tides are semi-diurnal with a neap tidal range of 1.27 m and a spring tidal range of 1.65 m. The annual averaged deep water significant wave height, as determined from measurements, is 0.8 m (Hicks and Hume, 1997). The spring tidal prism is 96×10^6 m^3. The throat cross-sectional area is 4,680 m^2. From this, the maximum cross-sectionally averaged velocity at spring tide is 1.43 m s^{-1}. The mean throat depth is 12 m (Hume and Herdendorf, 1992).

The inlet is bounded by a rocky headland on the north and a barrier island to the south. The ebb delta has a volume of 30×10^6 m^3, extending approximately 3 km offshore. A well-defined channel connects inlet and ocean. The depth over the ebb delta at the distal end of the channel is 2.5 m. Various papers dealing with Katikati Inlet show rather different values for magnitude and direction of the longshore sand transport (Hicks and Hume, 1996, 1997; Hicks et al., 1999; Hume and Herdendorf,

Figure 4.6 Katikati Inlet, New Zealand, in 2015 (Source: Google Earth).

1992). Based on the numbers in these papers, the gross longshore sand transport is estimated to be less than 0.5×10^6 m^3 year^{-1}.

Hicks et al. (1999), based on personal communication with T.M. Hume, concluded that the mode of bypassing at Katikati Inlet is tidal flow bypassing. Images of Google Earth, covering the period 2003–2013, show that the position of the main ebb channel has not changed much during this period and thus location stability is good. With a spring tidal prism of 96×10^6 m^3 and assuming a maximum gross longshore sand transport not exceeding 0.5×10^6 m^3 year^{-1}, the value of P/M is larger than 190.

4.8 Ameland Inlet

Ameland Inlet is one of the tidal inlets on the Dutch Wadden coast (Fig. 4.7). Tides are semi-diurnal with mean and spring tidal ranges of, respectively, 1.96 m and 2.26 m. The annual average deep water significant wave height is 1.10 m with a dominant northwest direction. The mean tidal prism is 434×10^6 m^3 and the spring tidal prism is 500×10^6 m^3. The inlet cross-sectional area is 27,780 m^2. Using these values the maximum cross-sectionally averaged velocity for mean tide conditions is 1.10 m s^{-1} and for spring tide conditions is 1.27 m s^{-1} (Israel and Dunsbergen, 1999; van de Kreeke, 1998).

The inlet is located between the barrier islands Terschelling to the west and Ameland to the east. The inlet connects the Wadden Sea to the North Sea. It has a

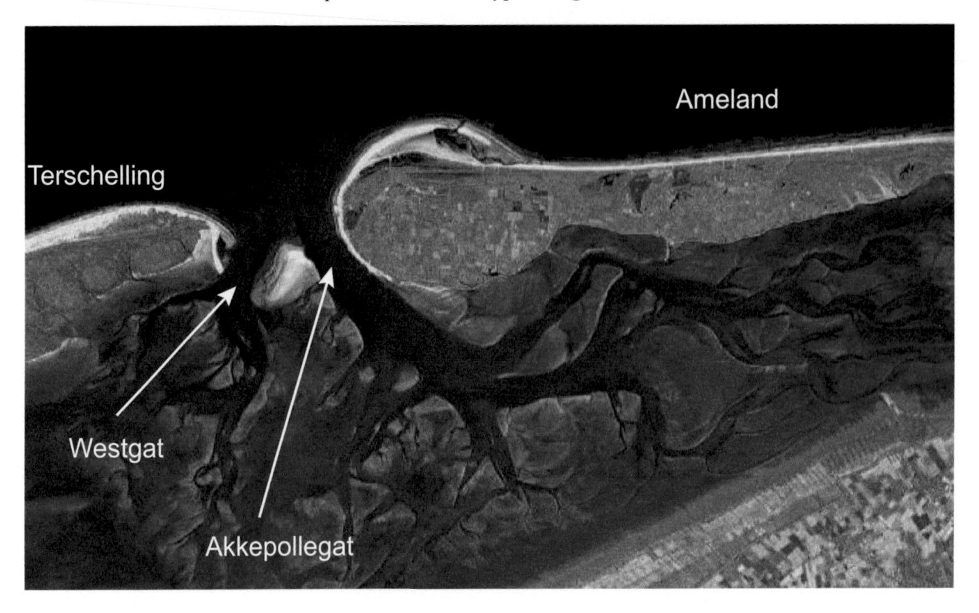

Figure 4.7 Ameland Inlet, The Netherlands, in 2010 (Esri et al., 2016).

stable throat, which can be attributed to channel and bank protection works along the southwest coast of Ameland. The average depth of the inlet throat is 15 m. In agreement with the large tidal prism, the Ameland Inlet has a large ebb delta with a volume of 130×10^6 m^3 (Louters and Gerritsen, 1994). The shape of the ebb delta is asymmetric which is attributed to the relatively strong, 0.5–1 m s^{-1}, shore-parallel tidal currents (Sha, 1989; van Veen, 1936). The estimated gross longshore sand transport is 0.7×10^6 m^3 year^{-1} (Cheung et al., 2007).

The ebb delta has two channels, Westgat to the west and the Akkepollegat to the east. The size and to a lesser extent also the orientation of these channels show a cyclic variation with a period of 50–60 years (Israel and Dunsbergen, 1999). This cycle is associated with the evolution of the inlet throat from a one-channel to a two-channel system. Throughout the morphological cycle Westgat varies between ebb dominant and flood dominant. Akkepollegat is always ebb-dominant. The wave-induced sand transport along the updrift Terschelling coast upon reaching the inlet is temporarily stored on the Terschelling flat, located at the eastern tip of that island. Depending on the flood or ebb dominance of the Westgat, sand is carried from the Terschelling flat towards the gorge of the inlet or the distal end of the Westgat. The flow in Akkepollegat is responsible for the transport of the sand to the distal end of this channel where it forms a shoal. From here the sand is transported by waves onto the ebb delta platform where it forms large bar complexes. The onshore migration of the bar complexes is intermittent having a timescale of 50–60 years. Once it reaches the shore, part of the sand is transported back to the inlet and another part travels along the shore in an eastward direction. The sand traveling along the shore exerts itself in the form of a progressive attenuating sand wave (Cheung et al., 2007). For this type of sand wave reference is made to Section 3.6.2.

Based on the foregoing, it is concluded that the bypassing mode is tidal flow bypassing and location stability is good. With a value of the spring tidal prism of 500×10^6 m^3 and a value of the gross longshore sand transport of 0.7×10^6 m^3 year^{-1}, the ratio $P/M = 714$.

5

Empirical Relationships

5.1 Introduction

For inlets in a sandy environment, relationships exist between parameters characterizing morphology and water motion. The more well known are the relationship between the cross-sectional area and tidal prism and the relationship between the ebb delta volume and the tidal prism. Attempts to correlate width and depth of the inlet cross-section with the tidal prism are only moderately successful.

5.2 Cross-Sectional Area – Tidal Prism Relationship

5.2.1 Observations

Observations suggest a relationship between cross-sectional area (A) and tidal prism (P). The cross-sectional area is measured with respect to MSL at the location of the gorge or throat. Usually, the tidal prism pertains to spring tide conditions. The most common presentation of the A–P relationship is in the form of a power function,

$$A = CP^q, \tag{5.1}$$

with C and q constants to be determined from observations. The relationship implies that the cross-sectional area is in equilibrium with the hydraulic environment. This equilibrium is dynamic rather than static, i.e., the cross-sectional area, and to a lesser extent the tidal prism, oscillates about an annual mean value (Section 2.3). Unfortunately, annual mean values are seldom available as most A and P values pertain to observations on a given day. These daily values can differ considerably from the annual mean. For example, Byrne et al. (1974) reported spring–neap tide variations in the cross-sectional area of Wachapreague Inlet (VA) of 10 percent. Similarly, FitzGerald and Nummedal (1983) reported that Price Inlet

(SC) experienced a change in throat cross-sectional area of 8 percent during a single tidal cycle and 26 percent during a three-year period.

Using Eq. (5.1) as the regression equation, regression analysis has been applied to many data sets. Examples for US inlets include LeConte (1905), O'Brien(1931, 1969), Jarrett (1976), van de Kreeke (1992), and Powell et al. (2006). Results for Japanese inlets are presented in Shigemura (1980). For the Dutch and German Wadden Sea inlets reference is made to van de Kreeke (1998) and (Dieckmann et al., 1988), respectively. Townend (2005) presents results for UK inlets and Heath (1975) and Hume and Herdendorf (1992) for New Zealand inlets. Using the metric system, C-values for the different sets of inlets vary between 10^{-6} and 10^{-3}. Values of the exponent q are close to 1, varying in a narrow band between 0.80 and 1.05. There are several reasons why C and q values differ for different sets of inlets. Among others, they result from differences in longshore sand transport, grain size and density and tide characteristics. This will be further discussed in Section 5.2.2. In addition, indications are that values of C and q depend on the type of least square analysis. The least square analysis is carried out by either solving a nonlinear least square equation or by taking the log on both sides of Eq. (5.1) and applying a linear regression.

Instead of using a power function, O'Brien (1969) showed that for a set of eight natural inlets (no jetties) in the US, the A–P relationship is reasonably represented by the linear equation

$$A = C_l P, \tag{5.2}$$

where $C_l = 6.5 \times 10^{-5}\ \mathrm{m}^{-1}$ and correlation coefficient $r^2 = 0.99$. In addition to being simpler than Eq. (5.1), this equation has the advantage that it is dimensionally correct.

5.2.2 Physical Justification of the A–P Relationship

Physical justifications for the A–P relationship have been presented by a number of investigators (Kraus, 1998; Suprijo and Mano, 2004; van de Kreeke, 1998, 2004). Except for minor differences, the approach by each of these investigators is the same. The basic premise is that when at equilibrium, on an annual average base, a balance exists between the volume of sand entering the inlet on the flood and the volume of sand leaving the inlet on the ebb. Only a negligible volume of sand is assumed to be deposited in the basin. The sand balance is assumed to hold for normal wave conditions, i.e., excluding storms.

Sand is carried towards the inlet by the wave-induced longshore sand transport and enters the inlet on the flood. For normal wave conditions, the volume of sand entering the inlet is taken as a fraction M' of the gross longshore sand transport M ($\mathrm{m}^3\ \mathrm{s}^{-1}$), i.e.,

$$Tr_{\text{fl}} = M', \tag{5.3}$$

where Tr_{fl} is sand transport into the inlet during normal wave conditions. For the sand transport into the inlet during storm conditions reference is made to Section 8.2 and 8.3.

For a physical justification of the A–P relationship given by Eq. (5.2), the sand transport leaving the inlet is taken in proportion to the power n of the velocity amplitude $\hat{u}$.

$$Tr_{\text{ebb}} = k\hat{u}^n, \tag{5.4}$$

where Tr_{ebb} is sand transport leaving the inlet and k is a coefficient depending on grain characteristics. Values of n are assumed to be between 3 and 5. Underlying assumptions are that the velocity can be reasonably approximated by a sine curve and velocities are considerably larger than the critical velocity of erosion over most of the ebb cycle.

When the inlet is at equilibrium

$$k\hat{u}^n = M'. \tag{5.5}$$

With the velocity not exactly sinusoidal, the velocity amplitude is defined as

$$\hat{u} = \frac{\pi P}{AT}, \tag{5.6}$$

where P is tidal prism, A is cross-sectional area and T is tidal period (Sorensen, 1977; van de Kreeke, 2004).

Substituting for $\hat{u}$ from Eq. (5.6) in Eq. (5.5) results in Eq. (5.2) with

$$C_l = \sqrt[n]{\frac{k}{M'}}\frac{\pi}{T}. \tag{5.7}$$

With M' being a fraction of M, the constant C_l decreases with increasing values of gross longshore sand transport.

The physical justification for the A–P relationship given by Eq. (5.1) follows along the same lines as the justification for Eq. (5.2), with the exception of the formulation of the ebb transport. Instead of Eq. (5.4), the ebb transport is written as

$$Tr_{\text{ebb}} = k\hat{u}^n W, \tag{5.8}$$

where W is the width of the inlet at MSL. Assuming that cross-sections with different cross-sectional area are geometrically similar, the width of the inlet is proportional to the square root of the cross-sectional area, i.e., $W = \beta_1\sqrt{A}$ (Appendix 8B). Substituting for W in Eq. (5.8) results in the expression for the ebb transport

$$Tr_{\text{ebb}} = k\hat{u}^n\sqrt{A}, \tag{5.9}$$

where the proportionality coefficient β_1 is incorporated in k. When the inlet is at equilibrium the ebb transport equals the flood transport, i.e.,

$$k\hat{u}^n\sqrt{A} = M'. \tag{5.10}$$

Substituting for $\hat{u}$ from Eq. (5.6) in Eq. (5.10) results in Eq. (5.1) with

$$C = \left(\frac{k\left(\frac{\pi}{T}\right)^n}{M'}\right)^{\frac{1}{n-\frac{1}{2}}} \quad \text{and} \quad q = \frac{n}{n-\frac{1}{2}}. \tag{5.11}$$

With M' being a fraction of M, C decreases with increasing values of the gross longshore sand transport. With $3 \leq n \leq 5$, values of q range from 1.11 to 1.20. For comparison, observed values of q range from 0.80 to 1.05, suggesting that the present physical justification is an oversimplification of the real physics. More research is needed to resolve this.

Even though some of the assumptions can be questioned, the physical justification for Eqs. (5.1) and (5.2) suggests that A–P relationships should only be expected to hold for sets of inlets that are geologically and hydrodynamically similar, i.e., inlets have the same grain characteristics, longshore sand transport and tidal period. Ocean tidal amplitudes, inlet lengths and basin surface areas can differ. Furthermore, for Eq. (5.1) to hold, cross-sections in the data set have to be geometrically similar. It follows that the coefficients C_l, C and q are not universal constants but are expected to differ from one geologically and hydrodynamically similar set of inlets to another. In this context it is of interest to note that Jarrett (1976) sampled a large number of US inlets and separated them into classes depending on the absence or presence of one or two jetties, reasoning that the jetties would affect the movement of sand into an inlet.

5.2.3 *Examples of* A–P *Relationships for Natural Inlets*

Data for tidal prisms and cross-sectional areas for five inlets on the Dutch Wadden Coast are presented in van de Kreeke (1998) and are reproduced in Table 5.1. The data pertains to mean tide conditions. The inlets are in a natural state (no jetties) and are scoured in fine to medium sand. Tides are semi-diurnal with offshore mean tidal ranges increasing from 1.60 m at Texel Inlet to 2.00 m at the Frisian Inlet (Dillingh, 2013). River flow is insignificant. The gross longshore sand transport is $0.5 - 1 \times 10^6$ m^3 year^{-1} (Spanhoff et al., 1997). Cross-sectional areas vary in a rather narrow range between 16,540 and 63,300 m^2. In view of similar tides, sediment characteristics and gross longshore sand transport, this set of natural inlets can be considered hydrodynamically and geologically similar.

Table 5.1 *Cross-sectional area* A *and tidal prism* P *for inlets of the Dutch Wadden Sea (van de Kreeke, 1998).*

Inlet	A [m^2]	P [10^6 m^3]
Texel Inlet	59,160	957
Eyerlands Gat Inlet	16,540	172
Vlie Inlet	63,300	848
Ameland Inlet	27,780	434
Frisian Inlet (before basin reduction)	24,540	321

P pertains to mean tide

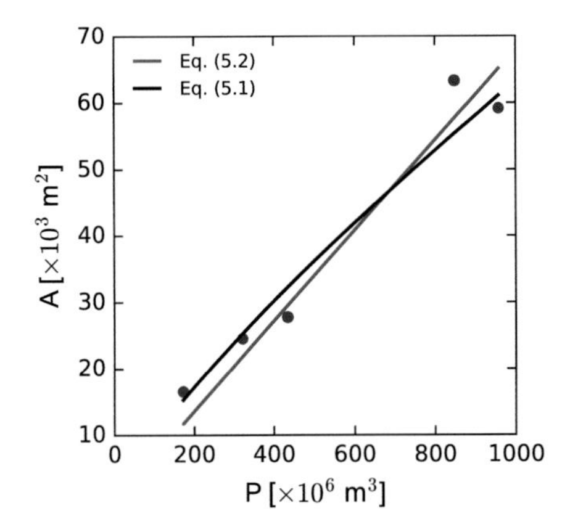

Figure 5.1 Cross-sectional area – tidal prism relationship for five inlets in the Dutch Wadden Sea for different regression equations. Using Eq. (5.1), $A = 3.4 \times 10^{-3} P^{0.81}$ with $r^2 = 0.96$. Using Eq. (5.2), $A = 6.8 \times 10^{-5} P$ with $r^2 = 0.95$.

A regression analysis for the five inlets shows the following results. Using the linear regression equation (5.2), $C_l = 6.8 \times 10^{-5}$ with $r^2 = 0.95$. Using the power function, Eq. (5.1), $C = 3.4 \times 10^{-3}$ and $q = 0.81$ with $r^2 = 0.96$. The trend lines for the two regression equations are presented in Fig. 5.1. Values of the correlation coefficients for the two regression equations are high and differ little. In spite of this, cross-sectional areas of individual inlets still can deviate substantially from the trend line. For example, using the linear regression equation and given the tidal prism, the cross-sectional area of the Eyerlandse Gat Inlet is ($6.8 \times 10^{-5} \times 172 \times 10^6 =$) 11,700 m^2, as opposed to 16,540 m^2 from observations. Possible reasons for the observations deviating from the trend line are discussed in Section 5.2.1.

Data for tidal prisms and cross-sectional areas for 16 inlets on the North Island of New Zealand are presented in Hume and Herdendorf (1992) and are reproduced in Table 5.2. The data pertains to spring tide conditions. The inlets are scoured in fine to medium sand. Tides are semi-diurnal with spring tidal ranges varying between 1.15 and 2.34 m. Inlets are located on a coast with a small longshore sand transport (Section 4.7). River flow is insignificant. Given that the wave climate is

Table 5.2 *Cross-sectional area* A *and tidal prism* P *for inlets on the North Island of New Zealand (Hume and Herdendorf, 1992).*

Inlet	A [m^2]	P [10^6 m^3]	Inlet	A [m^2]	P [10^6 m^3]
Whananaki	130	1.46	Whangapoua	980	8.54
Ngunguru	310	3.83	Whitianga	1,300	12.56
Pataua	140	2.24	Tairua	430	5.02
Whangarei	14,610	155	Whangamata	360	3.93
Mangawhai North	100	1.50	Puhoi	130	1.93
Mangawhai South	400	5.05	Maketu	70	0.79
Whangateau	660	10.53	Ohiwa	1,880	28.11
Katikati	4,680	95.82	Tauranga	6,260	130.8

P pertains to spring tide

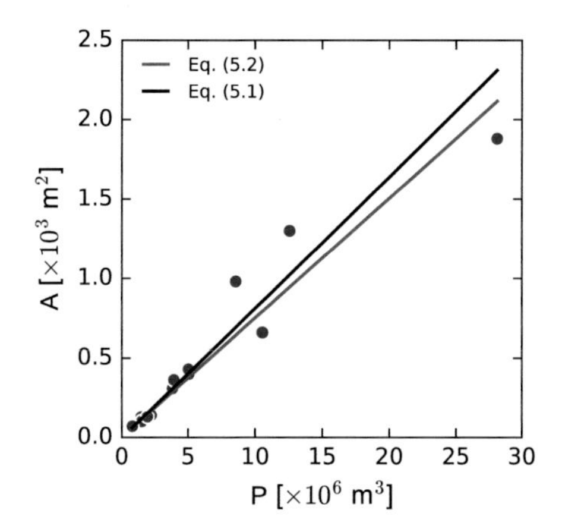

Figure 5.2 Cross-sectional area – tidal prism relationship for 13 inlets on the North Island of New Zealand for different regression equations. Using Eq. (5.1), $A = 6.5 \times 10^{-5} P^{1.01}$ with $r^2 = 0.96$. Using Eq. (5.2), $A = 7.5 \times 10^{-5} P$ with $r^2 = 0.91$.

more or less the same for all the inlets, it is reasonable to assume that this is also the case for the longshore sand transport. To a good approximation, the inlets in the data set are geologically and hydrodynamically similar.

A regression analysis for the 16 inlets shows the following results. Using the linear regression equation, Eq. (5.2), $C_l = 7 \times 10^{-5}$ m^{-1} with $r^2 = 0.875$. Using the power function, Eq. (5.1), $C = 1.68 \times 10^{-4}$ and $q = 0.95$ with $r^2 = 0.98$. Three of the inlets, Whangarei, Katikati and Tauranga, have cross-sectional areas and tidal prisms that are an order of magnitude larger than the other inlets. Omitting these inlets from the data set and using Eq. (5.2), $C_l = 7.5 \times 10^{-5}$ m^{-1} with $r^2 = 0.91$. Using Eq. (5.1), $C = 6.5 \times 10^{-5}$ and $q = 1.01$ with $r^2 = 0.96$. The trend lines for the 13 inlets are presented in Fig. 5.2. Similar to the inlets of the Wadden Sea, the cross-sectional areas of the individual inlets can differ considerably from the trend line.

A direct comparison of the C, q and C_l values for the data sets of the Wadden Sea and New Zealand inlets is difficult because of the difference in the longshore sand transport and because the A–P relationships for the Wadden Sea inlets pertains to mean tide conditions and that for the New Zealand inlets pertains to spring tide conditions.

5.2.4 Equilibrium Velocity

In the equations for the sand balance, Eqs. (5.5) and (5.10), the value of the velocity amplitude $\hat{u}$ pertains to equilibrium conditions and is referred to as the equilibrium velocity $\hat{u}_{eq}$. For $\hat{u} > \hat{u}_{eq}$ the inlet erodes and for $\hat{u} < \hat{u}_{eq}$ the inlet shoals.

A difficulty in determining the value of $\hat{u}_{eq}$ from Eqs. (5.5) and (5.10) is ascertaining the values of k, n and M'. To circumvent this problem, recourse is taken to one of the A–P relationships discussed in Section 5.2.1.

With Eq. (5.1) as the A–P relationship and using Eq. (5.6) as the expression for the velocity, the equilibrium velocity is

$$\hat{u}_{eq} = \frac{\pi A^{\frac{1}{q}-1}}{T C^{\frac{1}{q}}}. \tag{5.12}$$

With values of q close to one, the equilibrium velocity is a weak increasing function of A for $q < 1$ and a weak decreasing function of A for $q > 1$. With Eq. (5.2) as the cross-sectional area – tidal prism relationship and using Eq. (5.6), the expression for the equilibrium velocity is

$$\hat{u}_{eq} = \frac{\pi}{C_l T}, \tag{5.13}$$

In this case the equilibrium velocity is independent of the cross-sectional area. With C and C_l decreasing for increasing values of the gross longshore sand transport, $\hat{u}_{eq}$ increasing with increasing values of the gross longshore sand transport.

Using the results of the linear regression, the equilibrium velocities are calculated for the inlets of the Dutch Wadden Sea and New Zealand. For the Wadden Sea inlets with $C_l = 6.8 \times 10^{-5}$ m^{-1} and $T = 44{,}712$ s, it follows from Eq. (5.13) that the equilibrium velocity $\hat{u}_{eq} = 1.03$ m s^{-1}. Similarly, for the 13 New Zealand inlets, with $C_l = 7.5 \times 10^{-5}$ m^{-1}, $\hat{u}_{eq} = 0.94$ m s^{-1}.

5.3 Relationship between Depth and Width of the Cross-Section and Tidal Prism

Attempts to correlate the width and depth of the inlet throat cross-section with the tidal prism are reported in Hume and Herdendorf (1992). Using data for a set of inlets on the North Island of New Zealand they conclude that: "While the cross-sectional area shows a strong correlation with the tidal prism, there is a much weaker correlation between depth and tidal prism ($r^2 = 0.82$) and width and tidal prism ($r^2 = 0.63$)." Apparently, the inlet width and depth are free to adjust while the cross-sectional area is controlled by the tidal prism. Other studies, e.g., Bruun (1981) and Marino and Mehta (1988), show similar results with width to mean depth ratios varying widely and no obvious relationship with cross-sectional area or tidal prism.

5.4 Ebb Delta Volume – Tidal Prism Relationship

Similar to the cross-sectional area of the inlet, the ebb delta volume, V, appears to maintain a consistent relationship with the tidal prism P. In terms of a power function,

$$V = aP^b. \tag{5.14}$$

This relationship is empirical and physical justification is based on the idea that, when the ebb jet disperses and its competency decreases, sand is deposited in the nearshore, thereby building up an ebb delta platform. As this platform accretes vertically, wave energy augments the flood tidal currents. The equilibrium volume is reached when the flood tidal and wave generated currents transport the same volume of sand onshore as the ebb currents transports offshore (Fitzgerald et al., 1984).

Based on observations and a regression analysis, values of the coefficients a and b are presented in Walton and Adams (1976). They distinguish between highly, moderately and mildly exposed coasts taking the parameter H^2T^2 as a measure of wave energy. H and T are average values of wave height and period, respectively, determined from wave gages located in the near-shore zone. Restricting attention to the US coast, examples of mildly exposed coasts are the South Carolina, Texas and lower Florida Gulf coasts. The east coast and Panhandle of Florida are in the moderate range and the Pacific coast is in the highly exposed range. For each type of coast the regression analysis showed reasonable correlation with values of $b = 1.08$ for the moderately, $b = 1.23$ for the highly and $b = 1.24$ for the mildly exposed coast. A second regression analysis was then carried out taking $b = 1.23$ for all inlet groupings. Using the metric system, the resulting values of the coefficient a, together with the number of inlets for each type of coast, are listed in Table 5.3. It is not clear in the analysis of Walton and Adams (1976) whether values of the tidal prism refer to spring or mean tide conditions. Information on correlation coefficients is not available. However, the graphs in Walton and Adams (1976), in which ebb delta volume is plotted versus tidal prism, show only

Table 5.3 *Results of the regression analysis performed by Walton and Adams (1976).*

Exposure	a	b	Number of inlets
Mild	10.12×10^{-3}	1.23	16
Moderate	7.70×10^{-3}	1.23	18
High	6.38×10^{-3}	1.23	7

Table 5.4 *Results of the regression analysis performed by Powell et al. (2006).*

Location	a	b	Number of inlets	r^2
Florida Atlantic coast	3.80×10^{-3}	1.26	28	0.57
Florida Gulf coast	1.41×10^{1}	0.73	39	0.70

a limited goodness of fit, as some inlets deviate substantially from the regression equation.

With b positive, the volume of the ebb delta increases with increasing values of the tidal prism. The coefficient a, and therefore the ebb delta volume, decreases, going from mildly to highly exposed coasts; more sand is stored in the ebb delta of a low-energy coast than in the ebb delta of a high-energy coast. This is in agreement with what has been stated in Section 2.4.2 when discussing the geomorphology of the ebb delta.

Powell et al. (2006), using Eq. (5.14), correlated the ebb delta volumes and tidal prisms for 28 inlets on the Florida Atlantic coast and 39 inlets on the Florida Gulf coast. They found values for the coefficients a and b, using the metric system, listed in Table 5.4. Similar to Walton and Adams (1976), with b positive, the ebb delta volume increases with increasing values of the tidal prism. In the same table, the number of inlets and the values of the correlation coefficients are indicated. In their data set they used the spring tidal prism. Values of correlation coefficients are low and, similar to the Walton and Adams (1976) regression analysis, it can be concluded that the goodness of fit is limited.

For the Florida Atlantic coast being moderately exposed, it seems justified to compare the results of Powell et al. (2006) with the results of the moderately exposed coast of Walton and Adams (1976). For Powell et al. (2006), the value of the exponent b is larger and the value of the coefficient a is smaller than for Walton and Adams (1976). As a result, given a value for the tidal prism, the calculated ebb delta volumes do not differ all that much. This is demonstrated for the smallest and largest value of the tidal prism in the Powell et al. (2006) data set, respectively, 2×10^5 m^3 and 3.3×10^8 m^3. For the smallest value, Walton and

Adams (Eq. (5.14), with $a = 7.7 \times 10^{-3}$ and $b = 1.23$) yields an ebb delta volume of 0.26×10^5 m^3 and Powell et al. (Eq. (5.14), with $a = 3.8 \times 10^{-3}$ and $b = 1.26$) yields a value of 0.18×10^5 m^3. Taking the largest tidal prism in the data set, the corresponding ebb delta volumes are 2.31×10^8 m^3 for Walton and Adams and 2.05×10^8 m^3 for Powell et al. In spite of the differences in the coefficients and exponents, the delta volumes for the Walton and Adams and the Powell et al. empirical equations show reasonable agreement.

5.5 Flood Delta Volume – Tidal Prism Relationship

In addition to the ebb delta volume, Powell et al. (2006) looked for a relationship between flood delta volume and tidal prism for the 39 inlets along the Florida Gulf coast. Using Eq. (5.14) as the regression equation and using the metric system, this resulted in $a = 6.95 \times 10^3$ and $b = 0.37$. The scatter in the data is substantial as witnessed by the r^2 value of 0.38. According to FitzGerald (1988, 1996), the limited correlation between flood delta volume and tidal prism is related to the fact that the size and presence of flood deltas depend on the amount of open water space in the back-barrier lagoon. This can vary considerably from one type of back-barrier lagoon to the other; see Fig. 2.4.

6

Tidal Inlet Hydrodynamics; Excluding Depth Variations with Tidal Stage

6.1 Introduction

The dynamics of the flow in the inlet are described by the equation for uniform unsteady open channel flow. Variations in depth with tidal stage are neglected. The dynamic equation is complemented with a continuity condition that assumes a pumping mode for the back-barrier lagoon, i.e., the water level in the back-barrier lagoon fluctuates uniformly. Although these are simplifications, the advantage is that they allow relatively simple analytical solutions that are helpful in identifying mechanisms responsible for phenomena such as resonance, tidal choking and generation of (odd) overtides. As examples, analytical solutions by Keulegan (1951, 1967) and Mehta and Özsoy (1978) are presented. Results of the analytical solutions are applied to a representative inlet and compared with numerical results.

6.2 Inlet Schematization

The tidal inlet system is schematized to an inlet and a back-barrier lagoon (Fig. 6.1). The inlet connects the back-barrier lagoon and the ocean. Its geometry is simplified to a prismatic channel with diverging sections at both ends. The back-barrier lagoon is schematized to a basin with uniform depth. Referring to Chapter 2, in the real world inlets have varying widths and depths and back-barrier lagoons are characterized by tidal flats and marsh areas. Therefore, the schematization presented in Fig. 6.1 is only a rough representation of an actual inlet.

6.3 Governing Equations and Boundary Condition

6.3.1 Dimensional Equations

In deriving the governing equations, the major assumptions are 1) one-dimensional unsteady uniform flow in the inlet, 2) a uniformly fluctuating water level in the basin (pumping or Helmholz mode) and 3) negligible variations in cross-sectional

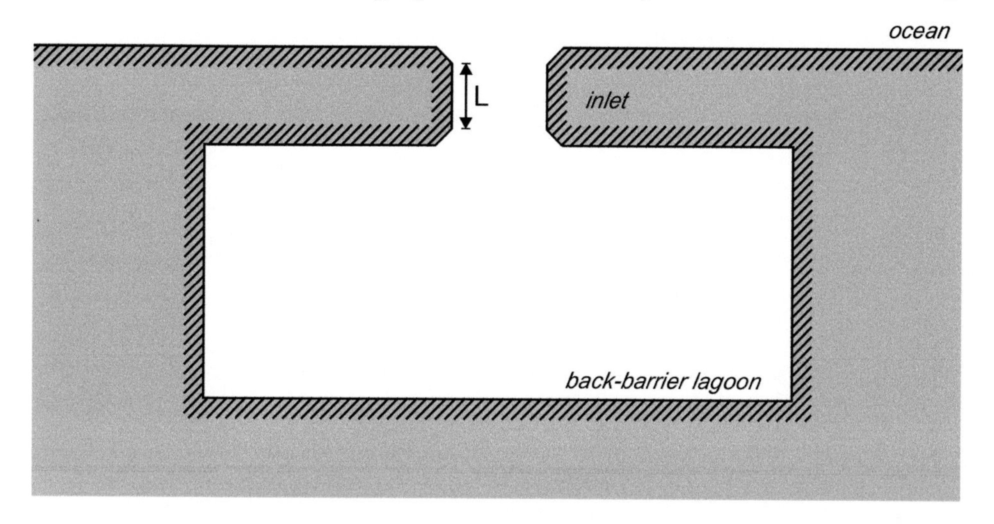

Figure 6.1 Inlet and back-barrier lagoon.

area of the inlet and basin surface area with tidal stage. With these assumptions, the equation for the flow in the inlet is (Appendix 6.A):

$$\frac{L}{g}\frac{du}{dt} + \left(\frac{m}{2g} + \frac{FL}{gR}\right) u|u| = \eta_0 - \eta_b. \tag{6.1}$$

In this equation u is the cross-sectionally averaged velocity, positive in the flood direction, L is length of the prismatic part of the inlet, g is gravity acceleration, t is time, $F = f/8$ where f is the Darcy–Weisbach friction factor, R is hydraulic radius, m is entrance/exit loss coefficient, η_0 is the ocean tide and η_b is the basin tide.

Eq. (6.1) is complemented by a continuity equation, expressing the equality of the water volume flux in the inlet and the rate of change of the basin volume. Assuming a uniformly fluctuating basin water level, this results in

$$uA = A_b \frac{d\eta_b}{dt}, \tag{6.2}$$

where A is cross-sectional area of the inlet channel, A_b is basin surface area and t is time. The ocean forcing is a simple harmonic tide

$$\eta_0 = \hat{\eta}_0 \sin \sigma t, \tag{6.3}$$

where $\hat{\eta}_0$ is the ocean tidal amplitude and σ is the angular frequency of the tide.

To illustrate the relative importance of the terms on the left-hand side of Eq. (6.1), the velocity is assumed to be simple harmonic,

$$u = \hat{u} \cos(\sigma t - \alpha). \tag{6.4}$$

Here $\hat{u}$ is the velocity amplitude and α is a phase angle. Using Eqs. (6.1) and (6.4), it follows that the order of magnitude values of the inertia, entrance/exit loss and bottom friction terms, respectively, the first, second and third term on the left-hand side of Eq. (6.1), are

$$\frac{\sigma L}{g}\hat{u}, \quad \frac{m}{2g}\hat{u}^2 \quad \text{and} \quad \frac{FL}{gR}\hat{u}^2.$$

Restricting attention to inlets in a sandy environment that have existed for some time, the velocity amplitude $\hat{u}$ is approximately 1 m s^{-1} and is referred to as the equilibrium velocity (see Section 5.2.4). A typical value of σ is 1.4×10^{-4} rad s^{-1}, corresponding with a semi-diurnal tide. From the literature, typical values of the coefficients m and F are, respectively, 1 and 3×10^{-3}. With these values it follows that the ratio of the friction and inertia term is approximately $20/R$. With R between 2 and 30 m (see Section 2.4.1), for the shallower inlets bottom friction is considerably larger than inertia. Similarly, the ratio of the bottom friction term and the entrance/exit loss term is approximately $6\times10^{-3}L/R$. Only for short and deep inlets, the value of the entrance/exit loss term approaches that of the bottom friction term.

6.3.2 Non-Dimensional Equations; Lumped Parameter Model

To reduce the number of parameters, Eqs. (6.1) and (6.2) are non-dimensionalized. Introducing the water level scale $\hat{\eta}_0$, velocity scale U and timescale σ^{-1}, the non-dimensional variables are

$$\eta_0^* = \frac{\eta_0}{\hat{\eta}_0}, \quad \eta_b^* = \frac{\eta_b}{\hat{\eta}_0}, \quad u^* = \frac{u}{U}, \quad t^* = \sigma t. \tag{6.5}$$

Using Eq. (6.2) the velocity scale is

$$U = \frac{\sigma \hat{\eta}_0 A_b}{A}. \tag{6.6}$$

Using the non-dimensional variables and the expression for U, the non-dimensional equations, corresponding to Eqs. (6.1) and (6.2), are

$$K_2^2 \frac{du^*}{dt^*} + \frac{1}{K_1^2} u^*|u^*| = \eta_0^* - \eta_b^*, \tag{6.7}$$

and

$$u^* = \frac{d\eta_b^*}{dt^*}. \tag{6.8}$$

The dimensionless parameters in Eq. (6.7) are defined as

$$K_1 = \frac{A}{\sigma \hat{\eta}_0 A_b}\sqrt{\frac{\hat{\eta}_0}{\left(\frac{m}{2g}+\frac{FL}{gR}\right)}}, \quad \text{and} \quad K_2 = \frac{\sigma}{\sqrt{\frac{gA}{LA_b}}}. \tag{6.9}$$

In terms of non-dimensional variables the forcing, Eq. (6.3), is

$$\eta_0^* = \sin(t^*). \tag{6.10}$$

The parameter K_1 is the Keulegan repletion factor and is the inverse of a damping factor. K_2 is the ratio of the forcing frequency σ and the natural or Helmholz frequency $\sqrt{gA/LA_b}$. The eight parameters, A_b, A, R, L, F, m, $\hat{\eta}_0$ and σ, in Eqs. (6.1) and (6.2) are lumped into the two non-dimensional parameters K_1 and K_2; hence the name *lumped parameter model* for Eqs. (6.7)–(6.10). For most natural inlets, $K_1 < 2$ (Seabergh, 2002).

6.4 Analytical Solution (Öszoy–Mehta)

6.4.1 Basin Tide and Inlet Velocity

In deriving a solution to Eqs. (6.7) and (6.8), Mehta and Özsoy (1978) assumed Eq. (6.7) to be weakly nonlinear. Together with the boundary condition given by Eq. (6.10), this suggests the following trial solutions for $\hat{\eta}_b^*$ and $\hat{u}^*$:

$$\eta_b^* = \hat{\eta}_b^* \sin(t^* - \alpha), \tag{6.11}$$

and

$$u^* = \hat{u}^* \sin(t^* - \beta). \tag{6.12}$$

Substituting the trial solutions in Eq. (6.8) results in

$$\hat{u}^* \sin(t^* - \beta) = \hat{\eta}_b^* cos(t^* - \alpha). \tag{6.13}$$

For this equation to hold requires that

$$\beta = \alpha - \frac{\pi}{2}, \tag{6.14}$$

and

$$\hat{u}^* = \hat{\eta}_b^*. \tag{6.15}$$

Substituting for β from Eq. (6.14) in Eq. (6.12) results in

$$u^* = \hat{u}^* \cos(t^* - \alpha). \tag{6.16}$$

In solving Eqs. (6.7) and (6.8), the nonlinear expression $u^*|u^*|$ is linearized by substituting for u^* from Eq. (6.16), i.e.,

$$u^*|u^*| = \hat{u}^{*2}\cos(t^* - \alpha)|\cos(t^* - \alpha)|. \tag{6.17}$$

Developing in a Fourier series, it follows that

$$u^*|u^*| = \frac{8}{3\pi}\hat{u}^{*2}\cos(t^* - \alpha) + \text{odd higher harmonics}. \tag{6.18}$$

Retaining only the first term in the series expansion and substituting in Eq. (6.7) results in

$$K_2^2\frac{du^*}{dt^*} + \frac{1}{K_1^2}\frac{8}{3\pi}\hat{u}^*u^* = \eta_0^* - \eta_b^*. \tag{6.19}$$

Substituting for η_0^*, η_b^* and u^* from, respectively, Eqs. (6.10), (6.11) and (6.16) in Eq. (6.19) and using of Eq. (6.15), it follows that

$$-K_2^2\hat{u}^*\sin(t^* - \alpha) + \frac{1}{K_1^2}\frac{8}{3\pi}\hat{u}^{*2}\cos(t^* - \alpha) = \sin(t^*) - \hat{u}^*\sin(t^* - \alpha). \tag{6.20}$$

Expanding the trigonometric functions and collecting terms proportional to $\sin(t^*)$ results in

$$-K_2^2\hat{u}^*\cos(\alpha) + \frac{1}{K_1^2}\frac{8}{3\pi}\hat{u}^{*2}\sin(\alpha) = 1 - \hat{u}^*\cos(\alpha). \tag{6.21}$$

Collecting terms proportional to $\cos(t^*)$ results in

$$K_2^2\hat{u}^*\sin(\alpha) + \frac{1}{K_1^2}\frac{8}{3\pi}\hat{u}^{*2}\cos(\alpha) = \hat{u}^*\sin(\alpha). \tag{6.22}$$

Multiplying Eq. (6.21) by $\cos(\alpha)$ and Eq. (6.22) by $\sin(\alpha)$ and subtracting, it follows that

$$\cos(\alpha) = (1 - K_2^2)\hat{u}^*. \tag{6.23}$$

Multiplying Eq. (6.21) by $\sin(\alpha)$ and Eq. (6.22) by $\cos(\alpha)$ and adding, it follows that

$$\sin(\alpha) = \frac{1}{K_1^2}\frac{8}{3\pi}\hat{u}^{*2}. \tag{6.24}$$

Solving for α and $\hat{u}^*$ from Eqs. (6.23) and (6.24) gives

$$\hat{u}^* = \sqrt{\frac{\left[\left(1 - K_2^2\right)^4 + \frac{4\left(\frac{8}{3\pi}\right)^2}{K_1^4}\right]^{1/2} - \left(1 - K_2^2\right)^2}{\frac{2\left(\frac{8}{3\pi}\right)^2}{K_1^4}}}, \tag{6.25}$$

and

$$\alpha = \tan^{-1}\left(\frac{\frac{8}{3\pi}\hat{u}^*}{K_1^2\left(1-K_2^2\right)}\right). \tag{6.26}$$

It follows from Eqs. (6.23) and (6.24) that for $K_2 < 1$, α is in the first quadrant and for $K_2 > 1$, α is in the second quadrant.

For many inlets the inertia term in the dynamic equation is small compared to the bottom friction term. In that case, neglecting the inertia term in Eq. (6.7), it follows from Eqs. (6.21) and (6.22) that

$$\hat{u}^* = \sqrt{\frac{\left[1+\frac{4\left(\frac{8}{3\pi}\right)^2}{K_1^4}\right]^{1/2}-1}{\frac{2\left(\frac{8}{3\pi}\right)^2}{K_1^4}}} \tag{6.27}$$

and

$$\alpha = \tan^{-1}\left(\frac{\frac{8}{3\pi}\hat{u}^*}{K_1^2}\right). \tag{6.28}$$

The same result is obtained by taking $K_2 = 0$ in Eqs. (6.25) and (6.26).

For applications where the bottom friction and entrance/exit loss can be neglected, i.e., $K_1 \to \infty$, it follows from Eqs. (6.21) and (6.22) that

$$\hat{u}^* = \frac{1}{1-K_2^2} \quad \text{and} \quad \alpha = 0 \quad \text{for} \quad K_2 < 1, \tag{6.29}$$

and

$$\hat{u}^* = \frac{-1}{1-K_2^2} \quad \text{and} \quad \alpha = \pi \quad \text{for} \quad K_2 > 1. \tag{6.30}$$

Note that the same result is not obtained by taking $K_1 = \infty$ in Eqs. (6.25) and (6.26).

6.4.2 Nature of the Solution; Resonance

As illustrated in Figs. 6.2a and 6.2b, the overall behavior of the Öszoy–Mehta Solution is similar to that of a mass–spring–dashpot system with resonance in the neighborhood of $K_2 = 1$. Depending on the values of K_1 and K_2, the basin tidal amplitude may be smaller or larger than the ocean tidal amplitude. For $K_2 = 0$, the basin tidal amplitude is always smaller than the ocean tidal amplitude. For a

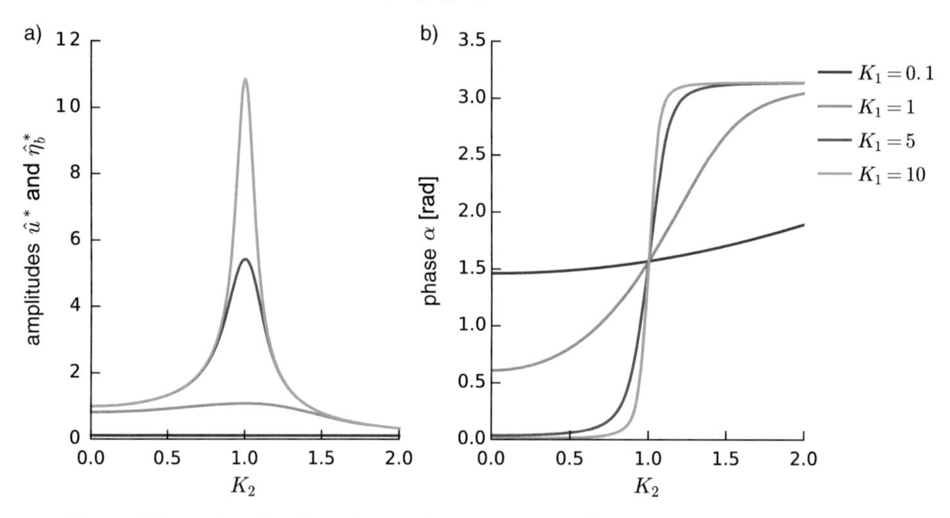

Figure 6.2 a) Amplitudes of non-dimensional basin tide and inlet velocity as a function of K_2 and K_1 and b) Phase of basin tide as a function of K_2 and K_1.

given K_2, the basin tidal amplitude and velocity amplitude increase with increasing values of K_1. For a given K_1, the basin tide and inlet velocity are maximal in the neighborhood of $K_2 = 1$. For values of K_1 smaller than approximately 0.1, the basin tidal amplitude is an order of magnitude smaller than the ocean tidal amplitude. These inlets have been referred to as choked inlets (Kjerfve, 1986).

Analytical solutions similar to the one presented by Mehta and Özsoy (1978) are found in Walton and Escoffier (1981), DiLorenzo (1988), Dean and Dalrymple (2002), Walton (2004b) and de Swart and Zimmerman (2009). Except for DiLorenzo (1988) and Walton (2004b), forcing is by a simple harmonic ocean tide. DiLorenzo (1988) included forcing by higher harmonics (overtides). In Walton (2004b) forcing is by a water level consisting of several astronomical components. Neglecting inertia, analytical solutions are presented in Brown (1928) and van de Kreeke (1967). These solutions agree with Eqs. (6.27) and (6.28).

6.5 Semi-Analytical Solution (Keulegan)

6.5.1 Basin Tide and Inlet Velocity

Keulegan (1951, 1967) derived expressions for the basin tide and the inlet velocity neglecting inertia. With $K_2 = 0$, the simplified set of non-dimensional equations is

$$\frac{1}{K_1^2} u^* |u^*| = \eta_0^* - \eta_b^* \tag{6.31}$$

and

$$u^* = \frac{d\eta_b^*}{dt^*}, \tag{6.32}$$

with the open boundary condition

$$\eta_0^* = \sin(t^*). \tag{6.33}$$

In the following, the major steps in solving for η_b^* and u^* are presented. For details, reference is made to Keulegan (1951, 1967).

It follows from Eqs. (6.31) and (6.32) that the basin and ocean tide curves intersect when the basin tide is maximal and minimal. At those times, the inlet velocity is zero. With this in mind, a qualitative plot of basin tide and ocean tide is presented in Fig. 6.3. In deriving a solution to Eqs. (6.31) and (6.32), Keulegan took the origin of the time axis at the low water intersection of the ocean and basin tide curve, resulting in the expression for the ocean tide

$$\eta_0^* = \sin(t^* - \tau). \tag{6.34}$$

The phase τ is defined in Fig. 6.3. Note that τ is an unknown quantity.

Eliminating the velocity u^* between Eqs. (6.31) and (6.32) results in the following equation for the basin tide:

$$\frac{1}{K_2^2}\frac{d\eta_b^*}{dt^*}\left|\frac{d\eta_b^*}{dt^*}\right| = \eta_0^* - \eta_b^*. \tag{6.35}$$

Rather than the basin tide, Keulegan took the difference between ocean tide and basin tide,

$$z^* = \eta_0^* - \eta_b^*, \tag{6.36}$$

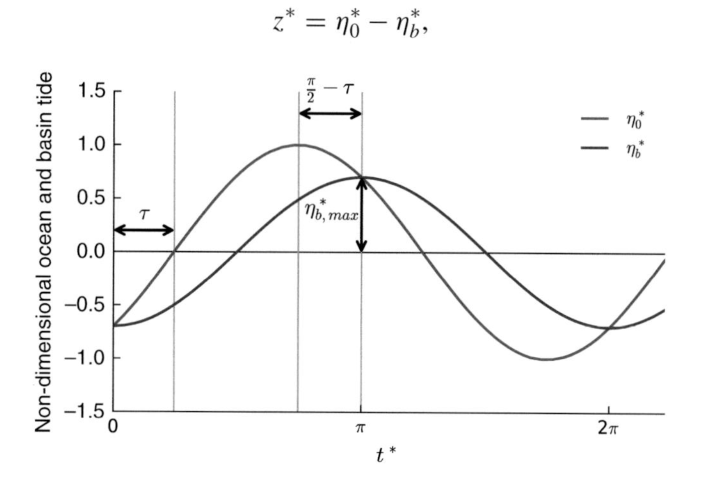

Figure 6.3 Ocean and basin tide curves; definition of time origin.

as the unknown. He assumed, and later verified, that for $0 < t^* < \pi$, z^* is positive and for $\pi < t^* < 2\pi$, z^* is negative. Limiting attention to the range $0 < t^* < \pi$ with $z^* > 0$ and η_0^* given by Eq. (6.34), it follows from Eq. (6.35) that

$$\frac{dz^*}{dt^*} = -K_1\sqrt{z^*} + \cos(t^* - \tau). \tag{6.37}$$

The corresponding equation for the range $\pi < t^* < 2\pi$ with $z^* > 0$ is

$$\frac{dz^*}{dt^*} = -K_1\sqrt{z^*} - \cos(t^* - \tau). \tag{6.38}$$

To solve for z^* from Eqs. (6.37) and (6.38), Keulegan showed that, in the interval $0 < t^* < 2\pi$, z^* the function $z^*(t^*)$ has zero mean and is asymmetric with the property

$$z^*(t^*) = -z^*\left(t^* + \tfrac{\pi}{2}\right). \tag{6.39}$$

This implies that, when developing $z^*(t^*)$ in a Fourier series, only odd harmonics are present. Limiting the Fourier series to a first and a third harmonic results in

$$z^*(t^*) = a_1 \sin(t^*) + a_1 b_3 \cos(t^*) - a_1 b_3 \cos(3t^*) - a_1 a_3 \sin(3t^*). \tag{6.40}$$

Substituting for $z^*(t^*)$ from Eq. (6.40) in Eq. (6.37) results in

$$\begin{aligned} &a_1 \cos(t^*) - a_1 b_3 \sin(t^*) + 3a_1 b_3 \sin(3t^*) - 3a_1 a_3 \cos(3t^*) - \cos(t^* - \tau) \\ &= -K_1\sqrt{a_1 \sin(t^*) + a_1 b_3 \cos(t^*) - a_1 b_3 \cos(3t^*) - a_1 a_3 \sin(3t^*)}. \end{aligned} \tag{6.41}$$

Developing the right-hand side of Eq. (6.41) in a Fourier series results in a series with only odd harmonics. With the right-hand side replaced by the first and third harmonic of the Fourier series, Eq. (6.41) contains terms in $\sin(t^*)$, $\cos(t^*)$, $\sin(3t^*)$ and $\cos(3t^*)$. Collecting terms proportional to each of these harmonics leads to four equations with unknowns a_1, a_3, b_3 and τ. Numerically solving these equations leads to values for a_1, a_3, b_3 and τ as functions of K_1. For K_1 between 0.1 and 5, values are presented in Table 6.1. In particular, the value of τ increases with increasing values of the repletion coefficient K_1 and thus, from Fig. 6.3, the phase between basin and ocean tide decreases for increasing values of K_1.

Referring to Keulegan (1951, 1967), substituting for z^* from Eq. (6.40) in Eq. (6.38) instead of Eq. (6.37) and developing in a Fourier series results in the same values for a_1, a_3, b_3 and τ listed in Table 6.1.

Expressions for the basin tide follow from Eq. (6.36), with $\hat{\eta}_0^*$ given by Eq. (6.34) and z^* given by (6.40), resulting in

$$\eta_b^* = \sin(t^* - \tau) - a_1 \sin(t^*) - a_1 b_3 \cos(t^*) + a_1 b_3 \cos(3t^*) + a_1 a_3 \sin(3t^*). \tag{6.42}$$

Using Eq. (6.32), the corresponding expression for the velocity is

$$u^* = \cos(t^* - \tau) - a_1 \cos(t^*) + a_1 b_3 \sin(t^*) - 3a_1 b_3 \sin(3t^*) + 3a_1 a_3 \cos(3t^*). \tag{6.43}$$

Table 6.1 *Coefficients in the Keulegan Solution.*

K_1	a_1	a_3	b_3	τ (rad)	$\cos\tau$	$\sin\tau$	C
0.1	0.9936	−0.0001	−0.0052	0.1161	0.99327	0.11580	0.8106
0.2	0.9745	−0.0004	−0.0106	0.2314	0.97334	0.22934	0.8116
0.3	0.9435	−0.0009	−0.0164	0.3456	0.94086	0.33874	0.8126
0.4	0.9020	−0.0017	−0.0220	0.4571	0.89735	0.44137	0.8153
0.5	0.8515	−0.0028	−0.0282	0.5656	0.84425	0.53593	0.8184
0.6	0.7942	−0.0043	−0.0347	0.6699	0.78386	0.62091	0.8225
0.7	0.7325	−0.0063	−0.0418	0.7691	0.71856	0.69549	0.8288
0.8	0.6689	−0.0089	−0.0495	0.8620	0.65091	0.75917	0.8344
0.9	0.5997	−0.0123	−0.0579	0.9553	0.57732	0.81649	0.8427
1.0	0.5451	−0.0165	−0.0664	1.0265	0.51783	0.85551	0.8522
1.2	0.4369	−0.0281	−0.0849	1.1599	0.39949	0.91676	0.8751
1.4	0.3489	−0.0448	−0.1038	1.2649	0.30119	0.95357	0.9016
1.6	0.2811	−0.0661	−0.1201	1.3443	0.22449	0.97446	0.9267
1.8	0.2294	−0.0910	−0.1327	1.4041	0.16588	0.98614	0.9484
2.0	0.1893	−0.1177	−0.1401	1.4489	0.12160	0.99258	0.9650
3.0	0.0883	−0.2207	−0.1187	1.5411	0.0295	0.9996	0.9950
4.0	0.0532	−0.2606	−0.0802	1.5608	0.0104	0.9999	0.9993
5.0	0.0323	−0.2740	−0.0532	1.5650	0.0057	1.0000	0.9994

Contrary to the Öszoy–Mehta Solution, the Keulegan Solution includes a third harmonic. This harmonic enters through the quadratic bottom friction.

6.5.2 Maximum Basin Level and Maximum Inlet Velocity

For many practical applications it is the maximum basin water level and maximum inlet velocity that are of interest. Referring to Fig. 6.3, the basin water level is maximum at $t^* = \pi$. From the same figure it follows that the basin half-tidal range is

$$\eta^*_{b_{max}} = \sin(\tau), \tag{6.44}$$

with τ a function of K_1. Values of $\sin(\tau)$ as a function of K_1 are presented in Table 6.1 for values of K_1 between 0.1 and 5.

It follows from Eq. (6.44) that the maximum basin half-tidal range is never larger than one and thus the maximum basin water level is at best equal to the maximum ocean water level. This differs from the Öszoy–Mehta Solution where, dependent on the ratio of the tidal and natural frequency, water levels in the basin can be larger than those in the ocean.

To arrive at an expression for the maximum velocity, Eq. (6.43) is differentiated with respect to t^*. Taking the derivative equal to zero, solving for t^* and substituting the value in Eq. (6.43), results in an expression for the maximum velocity as a function of the coefficients a_1, a_3, b_3 and τ. These coefficients are functions of K_1 and so is the maximum velocity. Keulegan wrote the relation between the maximum velocity and K_1 as

$$u^*_{max} = C \sin(\tau), \tag{6.45}$$

with both C and $\sin\tau$ functions of K_1. Values of C and $\sin\tau$ are presented in Table 6.1, for values of K_1 between 0.1 and 5.

6.5.3 Relative Contribution of the Third Harmonic

Using Eq. (6.43), the ratios of the velocity amplitudes of the third and first harmonic are calculated for $K_1 < 2$, values typical for most tidal inlets. The values of the coefficients a_1, a_3, b_3 and τ in Eq. (6.43) are listed in Table 6.1. From these calculations it followed that for $K_1 < 2$ the ratios of the velocity amplitudes remained practically the same with a value of approximately 0.12. The relative phase of the third harmonic, γ, as defined by Eq. (7.55) in Chapter 7, varied in a relatively narrow range between $3/4\pi$ and π. Using Fig. 7.2, it then follows that for this phase range the third harmonic lowers the peak velocity.

6.5.4 Multiple Inlets

The Keulegan Solution can be expanded to include multiple inlets, i.e., more than one inlet connecting the same basin to the ocean. For this an overall repletion coefficient,

$$K = \sum K_{1_j}, \tag{6.46}$$

is defined, with K_{1_j} the repletion coefficient of the jth inlet. The summation is over all the inlets. The overall repletion coefficient may be interpreted as the repletion coefficient of a single equivalent inlet, resulting in the same water level in the basin as the multiple inlets.

Proceeding in the same fashion as for the single inlet, the basin water level, η^*_b, and velocity, u^*, for the equivalent channel are calculated from Eqs. (6.42) and (6.43), respectively. Values of a_1, a_3, b_3 and τ are obtained from the Table 6.1, with K_1 replaced by K. Velocities for the individual inlets follow from

$$u^*_j = u^* \frac{K_{1_j}}{K}. \tag{6.47}$$

The basin half-tidal range, $\eta^*_{b_{\max}}$, and the maximum velocity for the equivalent channel, $u^*_{\max}$, are given by, respectively, Eqs. (6.44) and (6.45). In these equations

Table 6.2 *Typical parameter values for tidal inlets and parameter values for the representative inlet.*

Parameter	Symbol	Dimensions	Range	Value
Basin surface area	A_b	m^2	$0.5–5\times10^8$	1×10^8
Inlet cross-sectional area	A	m^2	$0.01–2\times10^4$	6×10^3
Inlet width	b	m	$0.05–2\times10^3$	1.5×10^3
Inlet depth	h	m	2–30	4
Inlet length	L	m	$0.5–5\times10^3$	3×10^3
Bottom friction factor	F	–	$2–4\times10^{-3}$	3.5×10^{-3}
Ocean tidal amplitude	$\hat{\eta}_0$	m	0.25–1.5	0.5
Tidal frequency	σ	rad s^{-1}	$0.7–1.4\times10^{-4}$	1.4×10^{-4}
Entrance/exit loss coefficient	m	–	0–2	1

Assumed is a rectangular cross-section with a width to depth ratio of 375, resulting in a shape factor $\beta_2 = 0.051$ (Appendix 8.A).

C and $\sin(\tau)$ are functions of K tabulated in Table 6.1, with K_1 replaced by K. Maximum velocities for the individual inlets follow from

$$u^*_{j\max} = u^*_{\max}\frac{K_{1_j}}{K}. \tag{6.48}$$

6.6 Application to a Representative Tidal Inlet

6.6.1 Representative Tidal Inlet

The Öszoy–Mehta and Keulegan Solutions are applied to a representative tidal inlet. The results are compared with a numerical solution. The planform and schematization of the representative tidal inlet are as indicated in Figs. 6.1 and 6.A.1. The selected dimensions and parameter values are typical for real world tidal inlets, and so are the ocean tidal amplitude and frequency. They are presented in the last column of Table 6.2. For comparison, ranges of dimensions and parameter values for real world tidal inlets are presented in column 4 of Table 6.2. A constraint in selecting the dimensions and parameter values of the representative tidal inlet is that the maximum cross-sectionally averaged velocity is approximately 1 m s^{-1}, a value encountered in many inlets located in a sandy environment.

6.6.2 Öszoy–Mehta Solution

The Öszoy–Mehta Solution is an approximate solution to Eqs. (6.7) and (6.8), with the ocean forcing given by Eq. (6.10). The solution is simple harmonic for both the basin tide and inlet velocity. With the parameter values in Table 6.2, the repletion coefficient for the representative inlet is $K_1 = 1.07$ and the natural or Helmholz frequency $K_2 = 0.32$.

The dimensional basin tide and inlet velocity are presented in Fig 6.4a together with the numerical solution. The numerical solution uses the same set of equations as the Öszoy–Mehta Solution. The equations are solved with an explicit time-staggered finite difference scheme (Reid and Bodine, 1969). For both solutions, the basin tide lags and the velocity leads the ocean tide. In contrast to the Öszoy–Mehta Solution, the numerical solution shows higher harmonics that are especially pronounced in the velocity. For the Öszoy–Mehta Solution, using Eq. (6.25), the velocity amplitude is 1.05 m s^{-1} and using Eqs. (6.25) and (6.15), the basin tidal amplitude is 0.44 m. For comparison, values of the numerically calculated maximum basin water level and maximum inlet velocity are 0.46 m and 0.94 m s^{-1}, respectively. It follows that the maximum basin water level for the Öszoy–Mehta Solution is somewhat lower and the maximum inlet velocity is significantly larger than for the numerical solution. Differences are attributed to the approximate nature of the Öszoy–Mehta Solution in which higher harmonics are neglected.

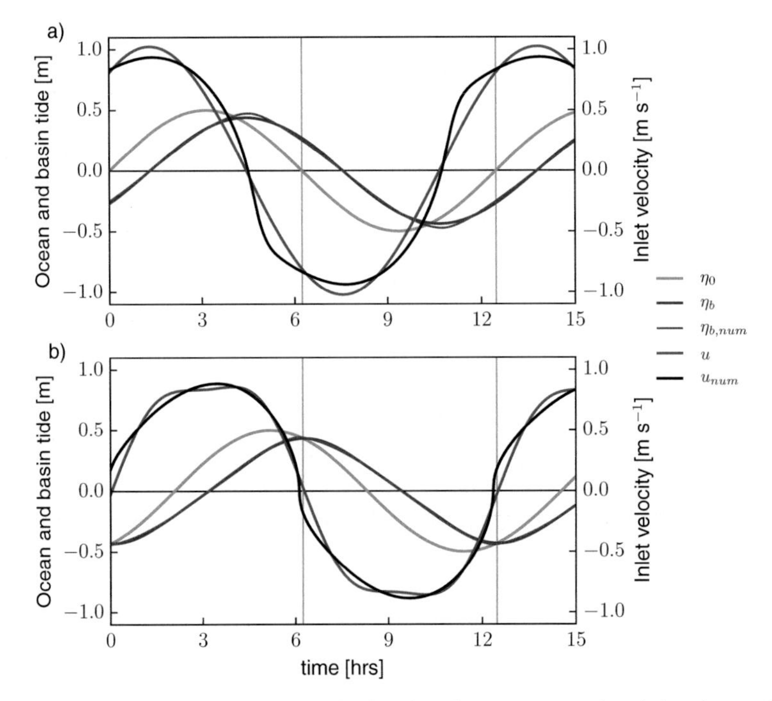

Figure 6.4 Basin tide and inlet velocity for the representative inlet determined from a) the Öszoy–Mehta Solution and b) the Keulegan Solution together with the ocean tide and numerical solutions. For parameter values, reference is made to Table 6.2.

6.6.3 Keulegan Solution

The Keulegan Solution is an approximate solution to Eqs. (6.31) and (6.32), with the ocean forcing given by Eq. (6.33). With the parameter values in Table 6.2, the repletion coefficient for the representative inlet $K_1 = 1.07$.

The dimensional basin tide and inlet velocity for the Keulegan Solution are presented in Fig. 6.4b, together with the numerical solution. The basin tide and inlet velocity are calculated from, respectively, Eqs. (6.42) and (6.43). The values of the coefficients a_1, a_3, b_3 and τ in these equations are determined from Table 6.1. The numerical solution uses the same set of equations as the Keulegan Solution. The equations are solved with an explicit time-staggered finite difference scheme. For both solutions, the basin tide lags and the inlet velocity leads the ocean tide. The ocean and basin tide curves intersect when the basin tide is maximum and minimum. At those times, velocities are zero. Using Table 6.1, it follows that $\sin(\tau) = 0.88$ and $C = 0.86$. Using Eq. (6.44), the maximum basin water level is 0.44 m and using Eq. (6.45), the maximum inlet velocity is 0.88 m s^{-1}. For comparison, values of the numerically calculated maximum basin water level and maximum inlet velocity are, respectively, 0.43 m and 0.89 m s^{-1}. Values of the Keulegan Solution are close to the same as those of the numerical solution.

The Keulegan Solution for the representative inlet consists of a first and third harmonic. The first and third harmonic for the inlet velocity are presented in Fig. 6.5. In agreement with what is stated in Section 6.5.3, the third harmonic lowers the peak velocity. Because there is no third harmonic in the ocean forcing, the third

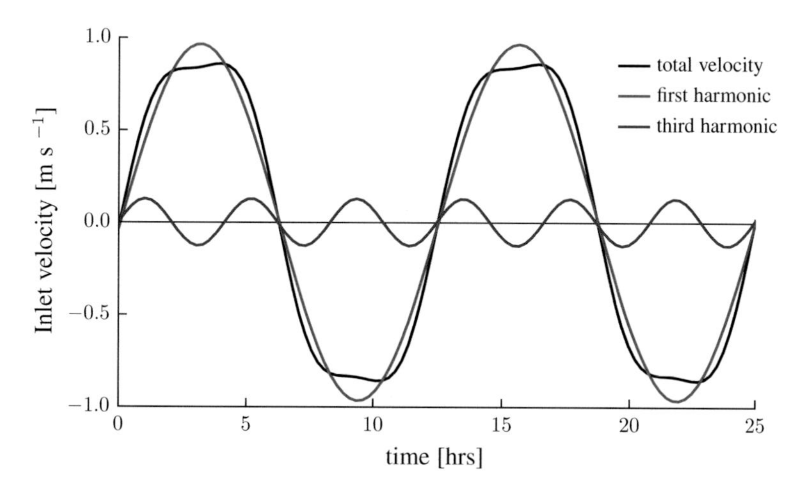

Figure 6.5 Contribution of the first and the third harmonic in the Keulegan Solution to the velocity of the representative inlet. For parameter values, reference is made to Table 6.2.

harmonic in the inlet velocity for the representative inlet is internally generated. The ratio of the velocity amplitudes of the third and first harmonic is 0.12. This seems a reasonable value; for example, the corresponding ratio for the Marsdiep Inlet (Texel Inlet), The Netherlands, as determined from long-term ADCP measurements, is 0.09. The difference can in part be attributed to the presence of the third harmonic in the ocean forcing of the Marsdiep Inlet (Buijsman and Ridderinkhof, 2007).

6.A Dynamics of the Flow in the Inlet

The schematized inlet consists of a prismatic part with diverging sections on both ends; see Fig. 6.A.1. In the prismatic part of the inlet, the flow dynamics is governed by the equations for unsteady uniform open channel flow (Dronkers, 1975), i.e.,

$$\text{for flood} \qquad \frac{L}{g}\frac{du}{dt} + \frac{FLu^2}{gR} = \eta_1 - \eta_2, \tag{6.A.1}$$

$$\text{for ebb} \qquad \frac{L}{g}\frac{du}{dt} - \frac{FLu^2}{gR} = \eta_1 - \eta_2. \tag{6.A.2}$$

Where u is the cross-sectionally averaged velocity, assumed to be independent of x, L is length of the prismatic part of the inlet, R is hydraulic radius, g is gravity acceleration, F is a friction factor, t is time and η_1 and η_2 are water levels at, respectively, x_1 and x_2; see Fig. 6.A.1. In the equations, variation in depth with tidal stage and hypsometric effects, the dependence of the surface area of the inlet on the water level, are neglected. For the equations including variation in depth with tidal stage reference is made to Chapter 7. Equations including hypsometric effects are presented in de Swart and Volp (2012).

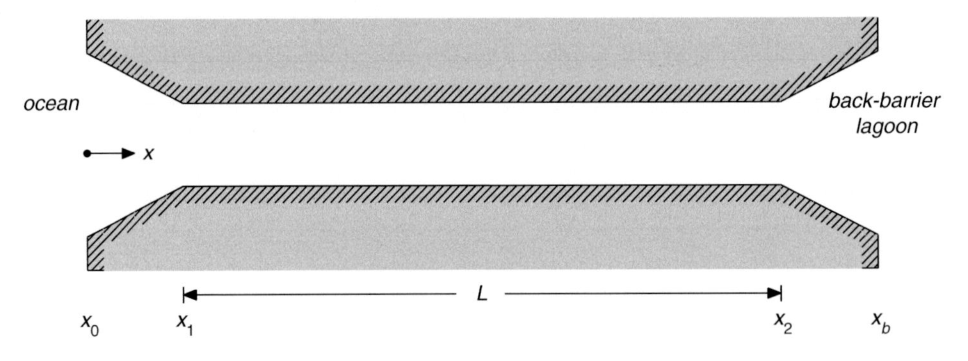

Figure 6.A.1 Schematization of the inlet.

The flow in the diverging sections constitutes a balance between advective accelerations and pressure gradients (Serrano et al., 2013; Vennell, 2006). Assuming u is independent of the cross-channel direction, this results in

$$u\frac{\partial u}{\partial x} + g\frac{\partial \eta}{\partial x} = 0. \tag{6.A.3}$$

For flood, integrating with respect to x between x_0 and x_1 and multiplying by the density ρ gives

$$\frac{1}{2}\rho u^2 + \rho g \eta_1 = \rho g \eta_0, \tag{6.A.4}$$

with η_0 is the ocean tide. At x_0, the cross-sectional area is relatively large, justifying the assumption of zero velocity. The left-hand side of Eq. (6.A.4) is the total energy (kinetic and potential) per unit volume at x_1. The right-hand side represents the potential energy per unit volume at x_0. When accounting for a turbulent kinetic energy loss, $\beta\rho u^2/2$ with $0 \leq \beta \leq 1$, Eq. (6.A.4) becomes

$$(1+\beta)\frac{1}{2}\rho u^2 + \rho g \eta_1 = \rho g \eta_0. \tag{6.A.5}$$

For flood, integrating with respect to x between x_2 and x_b and multiplying by the density ρ gives

$$\frac{1}{2}\rho u^2 + \rho g \eta_2 = \rho g \eta_b, \tag{6.A.6}$$

where η_b is the basin tide. At x_b the cross-sectional area is relatively large, justifying the assumption of zero velocity. The left-hand side of Eq. (6.A.6) represents the total energy per unit volume atx_2. The right-hand side of the equation is the potential energy per unit volume at x_b. When accounting for a turbulent kinetic energy loss, $\gamma\rho u^2/2$ with $0 \leq \gamma \leq 1$, Eq. (6.A.6) becomes

$$(1-\gamma)\frac{1}{2}\rho u^2 + \rho g \eta_2 = \rho g \eta_b. \tag{6.A.7}$$

Similar equations as (6.A.5) and (6.A.7) hold for the ebb. For the ocean side and basin side diverging section this results in, respectively,

$$(1-\gamma)\frac{1}{2}\rho u^2 + \rho g \eta_1 = \rho g \eta_0, \tag{6.A.8}$$

and

$$(1+\beta)\frac{1}{2}\rho u^2 + \rho g \eta_2 = \rho g \eta_b. \tag{6.A.9}$$

Solving for η_1 and η_2 from, respectively, Eqs. (6.A.5) and (6.A.7) and substituting in Eq. (6.A.1) results in

$$\frac{L}{g}\frac{du}{dt} + \left(\frac{m}{2g} + \frac{FL}{gR}\right)u^2 = \eta_0 - \eta_b, \tag{6.A.10}$$

with $m = \beta + \gamma$.

Solving for η_1 and η_2 from, respectively, Eqs. (6.A.8) and (6.A.9) and substituting in Eq. (6.A.2) results in

$$\frac{L}{g}\frac{du}{dt} - \left(\frac{m}{2g} + \frac{FL}{gR}\right)u^2 = \eta_0 - \eta_b. \tag{6.A.11}$$

With u positive in the flood direction and negative in the ebb direction, Eqs. (6.A.10) and (6.A.11) can be replaced by a single equation

$$\frac{L}{g}\frac{du}{dt} + \left(\frac{m}{2g} + \frac{FL}{gR}\right)u|u| = \eta_0 - \eta_b. \tag{6.A.12}$$

This equation corresponds to Eq. (6.1) in the main text.

7

Tidal Inlet Hydrodynamics; Including Depth Variations with Tidal Stage

7.1 Introduction

A difference with the previous chapter is that the hydrodynamic equations include depth variations with tidal stage. When forcing the tidal inlet system with a simple harmonic ocean tide, this results in a mean inlet velocity, a mean basin level different from that in the ocean and overtides (Boon and Byrne, 1981; LeProvost, 1991). A distinction is made between odd and even overtides. Odd overtides have frequencies that are odd multiples and even overtides have frequencies that are even multiples of the frequency of the ocean tide. Both odd and even overtides result in asymmetries in the inlet velocity and the basin tide (van de Kreeke and Robaczewska, 1993). It is the asymmetry in the inlet velocity introduced by even overtides that contributes to a long-term net sand transport (Pingree and Griffiths, 1979). For this reason, emphasis in this chapter is on even overtides, together with mean inlet velocity and mean basin level. The generation of the mean inlet velocity, the mean basin level and the even overtides is demonstrated by applying a perturbation analysis to the governing equations.

In addition to depth variations with tidal stage, variations in inlet width and basin surface area with tidal stage (sometimes referred to as hypsometric effects) will result in a mean inlet velocity, a mean basin level different from zero and even overtides. For this, reference is made to de Swart and Volp (2012) and Ridderinkhof et al. (2014).

7.2 Equations Including Depth Variations with Tidal Stage

Expanding Eqs. (6.1) and (6.2) to include depth variations with tidal stage results in

$$\frac{L}{g}\frac{du}{dt} + \left(\frac{m}{2g} + \frac{FL}{g(h+\eta_m)}\right)u|u| = \eta_0 - \eta_b, \qquad (7.1)$$

and

$$b(h + \eta_m)u = A_b \frac{d\eta_b}{dt}. \tag{7.2}$$

Assumed is a wide rectangular cross-section with mean depth h and width b. L is the length of the prismatic part of the inlet (Fig. 6.A.1), g is gravity acceleration, u is cross-sectional averaged velocity, positive in the flood direction, t is time, F is a nonlinear friction factor, m is an entrance/exit loss coefficient, η_0 is ocean tide, η_b is basin tide and A_b is basin surface area. The representative water level for the inlet, η_m, is defined as

$$\eta_m = \frac{\eta_0 + \eta_b}{2}. \tag{7.3}$$

To explain the generation of even overtides, mean inlet velocity and mean basin level, an analytical solution is sought to Eqs. (7.1) and (7.2). For this, the product $u|u|$ is linearized (Lorentz, 1926). Because the focus is on even rather than odd overtides, this is an acceptable simplification. After linearization, the term in Eq. (7.1) involving $u|u|$ is

$$\left(\frac{m}{2g} + \frac{FL}{g(h + \eta_m)}\right) u|u| = \left(\frac{m_l}{2g} + \frac{F_l L}{g(h + \eta_m)}\right) u, \tag{7.4}$$

with

$$m_l = \frac{8}{3\pi} \hat{u}_0 m \qquad \text{and} \qquad F_l = \frac{8}{3\pi} \hat{u}_0 F. \tag{7.5}$$

In Eq. (7.5), $\hat{u}_0$ is the leading-order velocity amplitude to be defined in Section 7.3, F_l is a linear friction factor and m_l is a linear entrance/exit loss coefficient. Using Eq. (7.4), Eq. (7.1) is written as

$$\frac{L}{g}\frac{du}{dt} + \left(\frac{m_l}{2g} + \frac{F_l L}{g(h + \eta_m)}\right) u = \eta_0 - \eta_b. \tag{7.6}$$

The ocean forcing is assumed to be a simple harmonic ocean tide

$$\eta_0 = \hat{\eta}_0 \sin \sigma t, \tag{7.7}$$

where $\hat{\eta}_0$ is the amplitude and σ is the angular frequency.

The nonlinear Eqs. (7.2) and (7.6) with the boundary condition, Eq. (7.7), are solved using a perturbation technique. The details of this method are described in Appendix 7.A. As shown in the appendix, the solution for u and η_b can be approximated by the sum of the solutions of a set of leading-order equations and a set of first-order equations.

The set of leading-order equations is

$$\frac{L}{g}\frac{du_0}{dt} + F' u_0 = \eta_0 - \eta_{b_0}, \tag{7.8}$$

$$A_b \frac{d\eta_{b_0}}{dt} - bhu_0 = 0, \tag{7.9}$$

with

$$F' = \frac{m_l}{2g} + \frac{F_l L}{gh}. \tag{7.10}$$

In these equations u_0 is the leading-order velocity and η_{b_0} is the leading-order basin water level. The set of first-order equations is

$$\frac{h^2}{F_l}\frac{du_1}{dt} + \left(\frac{m_l h^2}{2F_l L} + h\right) u_1 + \frac{gh^2}{F_l L}\eta_{b_1} = u_0 \eta_{m_0}, \tag{7.11}$$

$$\frac{A_b}{b}\frac{d\eta_{b_1}}{dt} - hu_1 = u_0 \eta_{m_0}, \tag{7.12}$$

with

$$\eta_{m_0} = \frac{\eta_0 + \eta_{b_0}}{2}. \tag{7.13}$$

In these equations, u_1 is the first-order velocity and η_{b_1} is the first-order basin water level. The leading-order term on the right-hand side of Eqs. (7.11) and (7.12) forces the first-order water motion and is the result of including variations of depth with tidal stage in the hydrodynamic equations. The complete solution consists of the sum of the leading- and first-order solutions.

7.3 Solution of the Leading-Order Equations

With the boundary condition given by Eq. (7.7), the following trial solutions are used to solve for η_{b_0} and u_0 from Eqs. (7.8) and (7.9),

$$\eta_{b_0} = \hat{\eta}_{b_0} \sin(\sigma t - \alpha), \tag{7.14}$$

and

$$u_0 = \hat{u}_0 \sin(\sigma t - \beta). \tag{7.15}$$

Substituting the trial solutions in Eq. (7.9) results in

$$A_b \sigma \hat{\eta}_{b_0} \cos(\sigma t - \alpha) = bh\hat{u}_0 \sin(\sigma t - \beta). \tag{7.16}$$

This equation can only be satisfied for

$$\beta = \alpha - \tfrac{\pi}{2}, \tag{7.17}$$

and

$$\hat{\eta}_{b_0} = \frac{bh}{\sigma A_b}\hat{u}_0. \tag{7.18}$$

Substituting for β from Eq. (7.17) in Eq. (7.15) results in

$$u_0 = \hat{u}_0 \cos(\sigma t - \alpha). \tag{7.19}$$

It follows from Eqs. (7.14) and (7.19) that the leading-order velocity and the leading-order basin tide are $\pi/2$ radians out of phase, with the velocity leading the basin tide.

Substituting the expression for η_{b_0} and u_0, respectively, Eqs. (7.14) and (7.19), in Eq. (7.8) and collecting terms proportional to $\sin \sigma t$ and $\cos \sigma t$ results in

$$\left(\hat{\eta}_{b_0} - \frac{\sigma L}{g} \hat{u}_0 \right) \cos \alpha + F' \hat{u}_0 \sin \alpha = \hat{\eta}_0, \tag{7.20}$$

and

$$-\left(\hat{\eta}_{b_0} - \frac{\sigma L}{g} \hat{u}_0 \right) \sin \alpha + F' \hat{u}_0 \cos \alpha = 0. \tag{7.21}$$

Eliminating $\hat{\eta}_{b_0}$ between Eqs. (7.18) and (7.21), and assuming $\hat{u}_0 \neq 0$, results in an equation for α,

$$-\left(1 - K_2^2\right) \sin \alpha + K' \cos \alpha = 0, \tag{7.22}$$

with

$$K' = F' \frac{\sigma A_b}{bh} \quad \text{and} \quad K_2 = \frac{\sigma}{\sqrt{\frac{gbh}{LA_b}}}. \tag{7.23}$$

Here, K' is a damping coefficient and K_2 is the ratio of the forcing and natural or Helmholtz frequency. It follows from Eq. (7.22) that

$$\tan \alpha = \frac{K'}{1 - K_2^2}. \tag{7.24}$$

For $K_2 < 1$, $\tan \alpha$ is positive and α is in the first or third quadrant and for $K_2 > 1$, $\tan \alpha$ is negative and α is in the second or fourth quadrant.

Substituting for $\hat{\eta}_{b_0}$ from Eq. (7.18) in Eq. (7.20) results in an equation for $\hat{u}_0$,

$$\left[\left(1 - K_2^2\right) \cos \alpha + K' \sin \alpha\right] \hat{u}_0 = \frac{\sigma A_b}{bh} \hat{\eta}_0. \tag{7.25}$$

For $K_2 < 1$ and α in the first or third quadrant, it follows from Eq. (7.25) that for $\hat{u}_0$ to be positive, α has to be in the first quadrant. Similarly, for $K_2 > 1$ with α in the second or fourth quadrant, it follows from Eq. (7.25) that for $\hat{u}_0$ to be positive, α has to be in the second quadrant.

Using Eq. (7.24), with $K_2 < 1$ and α in the first quadrant, the expressions for $\cos\alpha$ and $\sin\alpha$ are, respectively,

$$\cos\alpha = \frac{1 - K_2^2}{\sqrt{K'^2 + (1 - K_2^2)^2}}, \tag{7.26}$$

and

$$\sin\alpha = \frac{K'}{\sqrt{K'^2 + (1 - K_2^2)^2}}. \tag{7.27}$$

Using Eq. (7.24), the same expressions are found for $K_2 > 1$ and α in the second quadrant.

Substituting for $\cos\alpha$ and $\sin\alpha$ in Eq. (7.25) and solving for $\hat{u}_0$ results in

$$\hat{u}_0 = \frac{\sigma A_b}{bh} \frac{\hat{\eta}_0}{\sqrt{K'^2 + (1 - K_2^2)^2}}. \tag{7.28}$$

Substituting for $\hat{u}_0$ from Eq. (7.28) in Eq. (7.18) gives the expression for the leading-order basin tidal amplitude

$$\hat{\eta}_{b_0} = \frac{\hat{\eta}_0}{\sqrt{K'^2 + (1 - K_2^2)^2}}. \tag{7.29}$$

Because K', through Eqs. (7.5) and (7.10), depends on $\hat{u}_0$, evaluating $\hat{u}_0$ and $\hat{\eta}_{b_0}$ requires iterating. A first estimate of the value of $\hat{u}_0$ can be obtained using the Öszoy–Mehta Solution presented in Section 6.4.

7.4 Solution to the First-Order Equations

7.4.1 First-Order Forcing

The term on the right-hand side of Eqs. (7.11) and (7.12) constitutes the first-order forcing. By substituting the leading-order expressions η_0, η_{b_0} and u_0, given by Eqs. (7.7), (7.14) and (7.19), respectively, the first-order forcing term reads

$$u_0\eta_{m_0} = \tfrac{1}{4}\hat{\eta}_0\hat{u}_0\left[\sin(2\sigma t - \alpha) + \sin\alpha\right] + \tfrac{1}{4}\hat{\eta}_{b_0}\hat{u}_0\sin(2\sigma t - 2\alpha). \tag{7.30}$$

In this equation α, $\hat{u}_0$ and $\hat{\eta}_{b_0}$ are given by Eqs. (7.24), (7.28) and (7.29), respectively.

The forcing consists of harmonics with frequency 2σ and a time-independent part. As a result, the solution to the first-order equations, Eqs. (7.11) and (7.12), consists of a harmonic with frequency 2σ and a constant. For the velocity, this is written as

$$u_1 = \tilde{u}_1 + \langle u_1 \rangle, \tag{7.31}$$

and for the basin tide,

$$\eta_{b_1} = \tilde{\eta}_{b_1} + \langle \eta_{b_1} \rangle. \tag{7.32}$$

The tilde implies a harmonic with frequency 2σ, and thus $\tilde{u}_1$ and $\tilde{\eta}_{b_1}$ represent the first even overtides. The angle brackets denote tidal averaging, where $\langle u_1 \rangle$ is the mean velocity and $\langle \eta_{b_1} \rangle$ is the mean basin water level.

7.4.2 Mean Inlet Velocity and Mean Basin Level

Equations for the mean inlet velocity and mean basin level follow by tidally averaging the first-order equations, Eqs. (7.11) and (7.12). The resulting equations are

$$\left(\frac{m_l h^2}{2F_l L} + h \right) \langle u_1 \rangle + \frac{gh^2}{F_l L} \langle \eta_{b_1} \rangle = \langle u_0 \eta_{m_0} \rangle, \tag{7.33}$$

and

$$\langle u_0 \eta_{m_0} \rangle + h \langle u_1 \rangle = 0. \tag{7.34}$$

Using Eq. (7.30), the tidal average of the first-order forcing is

$$\langle u_0 \eta_{m_0} \rangle = \tfrac{1}{4} \hat{\eta}_0 \hat{u}_0 \sin \alpha. \tag{7.35}$$

Eliminating $\langle u_0 \eta_{m_0} \rangle$ between Eqs. (7.34) and (7.35) and using Eq. (7.27) results in the expression for the mean inlet velocity

$$\langle u_1 \rangle = -\frac{1}{4} \frac{A_b \sigma}{bh^2} \frac{\hat{\eta}_0^2 K'}{\left(K'^2 + (1 - K_2^2)^2 \right)}. \tag{7.36}$$

The mean inlet velocity is negative, i.e., in the ebb direction.

Eq. (7.34) expresses the balance between a tidally averaged water transport in the ebb direction and in the flood direction. The first term on the left represents a tidally averaged transport associated with the phase difference between the leading-order inlet tide and inlet velocity. It follows from Eq. (7.35) that, with $\sin \alpha$ positive (Section 7.3), this transport is in the flood direction. The second term on the left represents a tidally averaged transport in the ebb direction resulting from the mean inlet velocity.

Eliminating $\langle u_0 \eta_{m_0} \rangle$ between Eqs. (7.33) and (7.34) results in the expression for the mean basin level in terms of the mean inlet velocity

$$\langle \eta_{b_1} \rangle = -\left(\frac{m_l}{2g} + 2\frac{F_l L}{gh} \right) \langle u_1 \rangle. \tag{7.37}$$

Substituting for $\langle u_1 \rangle$ from Eq. (7.36), the mean basin level is

$$\langle \eta_{b_1} \rangle = \left(\frac{m_l}{8g} + \frac{F_l L}{2gh} \right) \frac{A_b \sigma}{bh^2} \frac{\hat{\eta}_0^2 K'}{\left(K'^2 + (1 - K_2^2)^2\right)}. \tag{7.38}$$

The mean basin level is positive and thus the basin level encounters a set-up.

For most tidal inlets the tidally averaged inlet velocity and basin tide are an order of magnitude smaller than the corresponding leading-order amplitudes. For example, using long-term measurements in the Frisian Inlet, van de Kreeke and Hibma (2005) report a leading-order velocity amplitude of 0.77 m s^{-1} and a tidally averaged velocity in the ebb direction of 0.05 m s^{-1}. The Frisian Inlet is a tidal inlet on the Dutch Wadden Sea coast; see Fig. 1.1a.

7.4.3 First-Order Tide and Velocity

Equations for the first-order tide follow by subtracting Eq. (7.33) from Eq. (7.11) and subtracting Eq. (7.34) from Eq. (7.12) resulting in, respectively,

$$\begin{aligned} \frac{h^2}{F_l} \frac{d\tilde{u}_1}{dt} + \left(\frac{m_l h^2}{2F_l L} + h \right) \tilde{u}_1 + \frac{gh^2}{F_l L} \tilde{\eta}_{b_1} &= \tfrac{1}{4} \hat{\eta}_0 \hat{u}_0 \sin(2\sigma t - \alpha) \\ &+ \tfrac{1}{4} \hat{\eta}_{b_0} \hat{u}_0 \sin(2\sigma t - 2\alpha), \end{aligned} \tag{7.39}$$

and

$$\frac{A_b}{b} \frac{d\tilde{\eta}_{b_1}}{dt} - h\tilde{u}_1 = \tfrac{1}{4} \hat{\eta}_0 \hat{u}_0 \sin(2\sigma t - \alpha) + \tfrac{1}{4} \hat{\eta}_{b_0} \hat{u}_0 \sin(2\sigma t - 2\alpha). \tag{7.40}$$

The following trial solutions for the unknowns $\tilde{\eta}_{b_1}$ and $\tilde{u}_1$ are proposed:

$$\tilde{\eta}_{b_1} = \hat{\hat{\eta}}_{b_1} \sin(2\sigma t - \varphi), \tag{7.41}$$

and

$$\tilde{u}_1 = \hat{\hat{u}}_1 \sin(2\sigma t - \psi). \tag{7.42}$$

Eq. (7.41) is written as

$$\tilde{\eta}_{b_1} = \hat{\hat{\eta}}_{b_{1s}} \sin 2\sigma t - \hat{\hat{\eta}}_{b_{1c}} \cos 2\sigma t, \tag{7.43}$$

with

$$\hat{\hat{\eta}}_{b_1} = \sqrt{\hat{\hat{\eta}}_{b_{1s}}^2 + \hat{\hat{\eta}}_{b_{1c}}^2}, \tag{7.44}$$

and

$$\tan \varphi = \frac{\hat{\hat{\eta}}_{b_{1c}}}{\hat{\hat{\eta}}_{b_{1s}}}. \tag{7.45}$$

Similarly, Eq. (7.42) is written as

$$\tilde{u}_1 = \hat{\tilde{u}}_{1s} \sin 2\sigma t - \hat{\tilde{u}}_{1c} \cos 2\sigma t, \tag{7.46}$$

with

$$\hat{\tilde{u}}_1 = \sqrt{\hat{\tilde{u}}_{1s}^2 + \hat{\tilde{u}}_{1c}^2}, \tag{7.47}$$

and

$$\tan \psi = \frac{\hat{\tilde{u}}_{1c}}{\hat{\tilde{u}}_{1s}}. \tag{7.48}$$

Substituting for $\tilde{\eta}_{b_1}$ and $\tilde{u}_1$ from, respectively, Eqs. (7.43) and (7.46) in Eqs. (7.39) and (7.40), and collecting terms in $\sin 2\sigma t$ and $\cos 2\sigma t$, results in four algebraic equations with unknowns $\hat{\tilde{\eta}}_{b_{1s}}$, $\hat{\tilde{\eta}}_{b_{1c}}$, $\hat{\tilde{u}}_{1s}$ and $\hat{\tilde{u}}_{1c}$. In matrix form:

$$\begin{bmatrix} 0 & \frac{gh^2}{F_l L} & \frac{2\sigma h^2}{F_l} & \left(\frac{m_l h^2}{2F_l L} + h\right) \\ -\frac{gh^2}{F_l L} & 0 & -\left(\frac{m_l h^2}{2F_l L} + h\right) & \frac{2\sigma h^2}{F_l} \\ \frac{2\sigma A_b}{b} & 0 & 0 & -h \\ 0 & \frac{2\sigma A_b}{b} & h & 0 \end{bmatrix} \begin{bmatrix} \hat{\tilde{\eta}}_{b_{1c}} \\ \hat{\tilde{\eta}}_{b_{1s}} \\ \hat{\tilde{u}}_{1c} \\ \hat{\tilde{u}}_{1s} \end{bmatrix} = \begin{bmatrix} F_c \\ F_s \\ F_c \\ F_s \end{bmatrix}, \tag{7.49}$$

with

$$F_c = -\tfrac{1}{4}(\hat{\eta}_0 \hat{u}_0 \cos \alpha + \hat{\eta}_{b_0} \hat{u}_0 \cos 2\alpha), \tag{7.50}$$

and

$$F_s = \tfrac{1}{4}(\hat{\eta}_0 \hat{u}_0 \sin \alpha + \hat{\eta}_{b_0} \hat{u}_0 \sin 2\alpha). \tag{7.51}$$

Solving for $\hat{\tilde{\eta}}_{b_{1s}}$, $\hat{\tilde{\eta}}_{b_{1c}}$, $\hat{\tilde{u}}_{1s}$ and $\hat{\tilde{u}}_{1c}$, the amplitude of the first-order basin tide and inlet velocity follow from Eqs. (7.44) and (7.47), respectively. The phase of the first-order basin tide and inlet velocity follows from Eqs. (7.45) and (7.48), respectively.

Solving the matrix equations (7.49) analytically is a cumbersome exercise. Instead, the system of equations is solved numerically using the parameter values of the representative inlet listed in Table 6.2. The results are presented in Section 7.6.

At most tidal inlets the amplitudes of the first-order velocity and basin tide are an order of magnitude smaller than the corresponding leading-order amplitudes. For example using long-term observations, for the Texel Inlet the ratio of the first and leading-order velocity amplitudes is 0.17 (Buijsman and Ridderinkhof, 2007) and for the Frisian Inlet is 0.08 (van de Kreeke and Hibma, 2005). Both inlets are part of the Dutch Wadden Sea coast; see Fig. 1.1a.

7.5 Tidal Asymmetry

Tidal asymmetry is associated with the combination of a fundamental tidal harmonic and its overtides. A distinction is made between water level asymmetry and velocity asymmetry. Here, emphasis is on velocity asymmetry because it results in a net sand transport (Pingree and Griffiths, 1979; van de Kreeke and Robaczewska, 1993).

To define tidal asymmetry, consider a rectilinear current velocity $u(t)$ that is periodic with zero mean. Defining the time origin at slack water, $u(t)$ is symmetric when

$$|u(t)| = |u(-t)|, \tag{7.52}$$

and is asymmetric when

$$|u(t)| \neq |u(-t)|. \tag{7.53}$$

Assuming the velocity consists of a fundamental harmonic and its first even overtide

$$u(t) = \hat{u}_0 \cos(\sigma t) + \hat{u}_1 \cos(2\sigma t - \beta). \tag{7.54}$$

In this equation, β is the phase of the even overtide with respect to the phase of the fundamental harmonic. It follows from Eqs. (7.52) and (7.53) that $u(t)$ is asymmetric except for $\beta = \pi/2$ and $\beta = 3\pi/2$. This is further demonstrated in Fig. 7.1. The asymmetry implies that for $-\pi/2 < \beta < \pi/2$ the maximum flood velocity (defined as positive) is larger than the maximum ebb velocity and the flood duration is shorter than the ebb duration. Assuming sand transport is proportional to $|u|^n \text{sign}(u)$, this leads to a long-term net sand transport in the flood direction. For $\pi/2 < \beta < 3\pi/2$, the maximum ebb velocity is larger than the maximum flood velocity, resulting in a net sand transport in the ebb direction (van de Kreeke and Robaczewska, 1993).

For a fundamental harmonic and its first odd overtide the expression for the velocity is

$$u(t) = \hat{u}_0 \cos(\sigma t) + \hat{u}_2 \cos(3\sigma t - \gamma). \tag{7.55}$$

In this equation, γ is the phase of the odd overtide relative to the phase of the first harmonic. Using Eqs. (7.52) and (7.53), it follows that the velocity $u(t)$ is asymmetric except for $\gamma = 0$ and $\gamma = \pi$. This is further demonstrated in Fig. 7.2. Contrary to the velocity asymmetry for the fundamental harmonic and its first even overtide, the velocity asymmetry for the fundamental harmonic and its first odd overtide does not lead to a net sand transport. The reason is that for the fundamental tidal harmonic and its first odd overtide holds that

$$u(t) = -u\left(t + \tfrac{T}{2}\right), \tag{7.56}$$

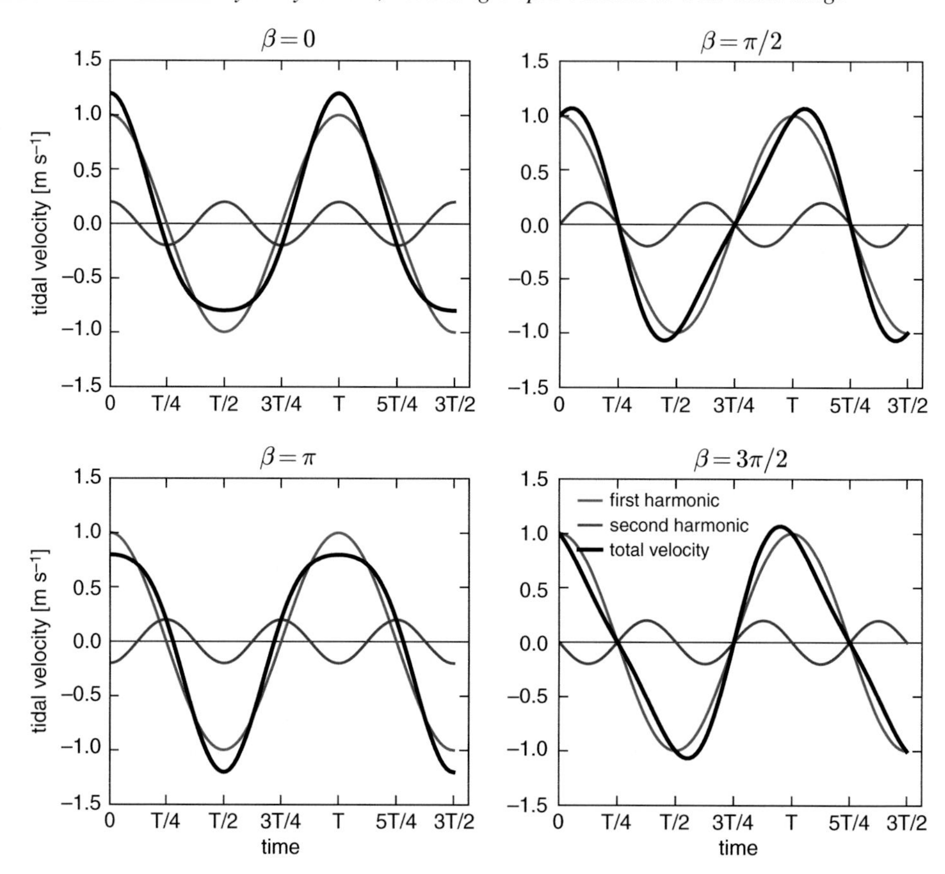

Figure 7.1 Leading-order tide, first even overtide and their sum for different values of the relative phase β. For definition of the symbols reference is made to Eq. (7.54) (adapted from van de Kreeke and Robaczewska, 1993).

with T is the period of the fundamental harmonic. Eq. (7.56) implies that regardless of the value of the relative phase γ, maximum flood velocities equal maximum ebb velocities.

For additional information on tidal asymmetry, reference is made to Boon and Byrne (1981), Dronkers (1986), Friedrichs and Aubrey (1988), Fry and Aubrey (1990), van de Kreeke and Dunsbergen (2000) and Dronkers (2005).

7.6 Application to the Representative Inlet

7.6.1 Leading-Order Solution

Parameter values for the representative inlet are presented in Table 6.2. Using this table, the value of K_2 defined by Eq. (7.23) is 0.32.

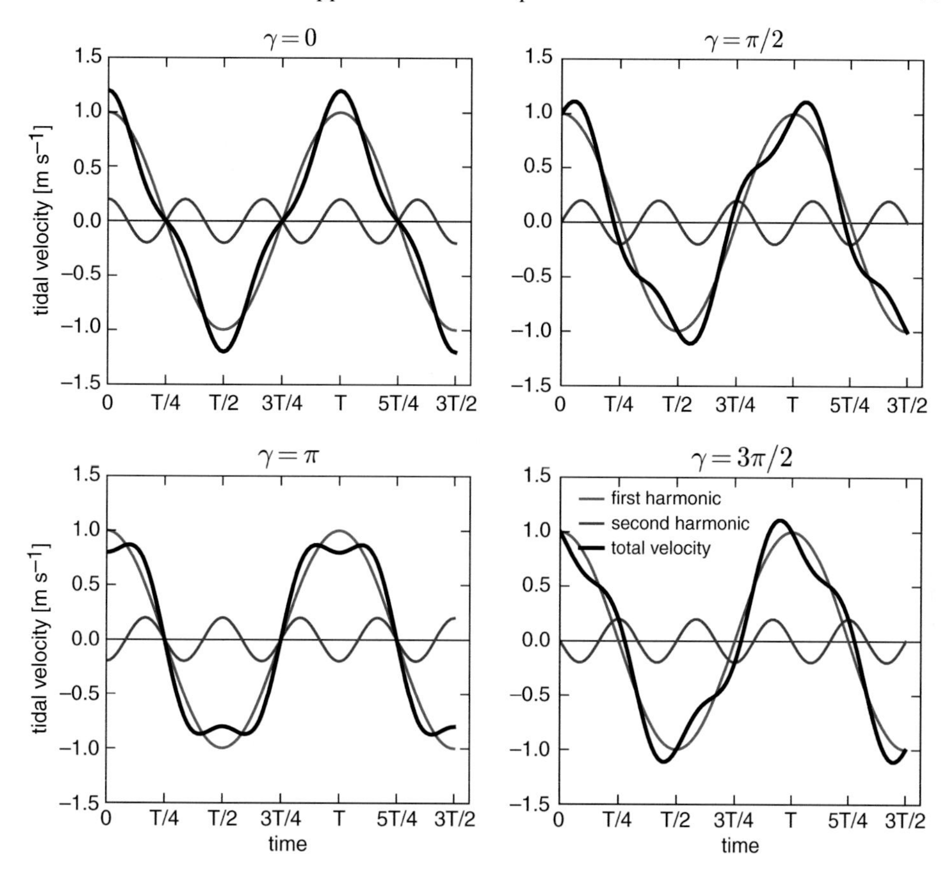

Figure 7.2 Leading-order tide, first odd overtide and their sum for different values of the relative phase γ. For definition of the symbols reference is made to Eq. (7.55) (adapted from van de Kreeke and Robaczewska, 1993).

Calculating the amplitudes and phases of the leading-order inlet velocity and basin tide requires an iterative approach. Calculations start with assuming a value for the velocity amplitude. The values of m_l and F_l are then calculated from Eq. (7.5). These values are used to calculate F' from Eq. (7.10). With the known value of F', K' is calculated from Eq. (7.23). With the known value of K', a new value of the velocity amplitude is calculated using Eq. (7.28). Calculations are carried out until the new and old velocity amplitudes are close to the same. A value of 1.05 m s^{-1} was taken as a first estimate of the amplitude of the leading-order velocity. This velocity amplitude was calculated using the Öszoy–Mehta Solution (Section 6.6.2). After iterating, the amplitude of the leading-order inlet velocity $\hat{u}_0 = 1.05$ m s^{-1}. The amplitude of the leading-order basin tide, calculated from Eq. (7.29), is $\hat{\eta}_{b_0} = 0.45$ m and using Eq. (7.24) the phase is $\alpha = 0.63$ rad. The values of the various coefficients are $m_l = 0.89$, $F_l = 3.12 \times 10^{-3}$, $F' = 0.28$ and $K' = 0.66$.

7.6.2 First-Order Solution

The first-order solution consists of the first even overtide of the velocity and basin tide and the tidally averaged inlet velocity and basin level. The amplitudes and phases of the overtides of the inlet velocity and basin level are numerically calculated using the matrix equation, Eq. (7.49). The results are $\hat{\hat{u}}_1 = 0.06$ m s^{-1}, $\psi = -0.23$ rad, $\hat{\hat{\eta}}_{b_1} = 0.02$ m and $\varphi = 1.91$ rad. The first-order amplitudes are an order of magnitude smaller than the corresponding leading-order amplitudes. Using Eq. (7.36), the mean velocity $\langle u_1 \rangle = -0.02$ m s^{-1} and from Eq. (7.37), the mean basin level $\langle \eta_{b_1} \rangle = 0.01$ m.

7.6.3 Tidal Asymmetry

Using Eqs. (7.19) and (7.42), the sum of the leading-order and first-order velocity is

$$u = \hat{u}_0 \cos(\sigma t - \alpha) + \hat{\hat{u}}_1 \sin(2\sigma t - \psi). \tag{7.57}$$

To determine the tidal asymmetry, i.e., ebb or flood dominance, this equation is written in the form of Eq. (7.54) by introducing the transformation of the time axis

$$\sigma t' = \sigma t - \alpha, \tag{7.58}$$

resulting in

$$u = \hat{u}_0 \cos(\sigma t') + \hat{\hat{u}}_1 \cos\left(2\sigma t' - \left(\left(\psi + \tfrac{\pi}{2}\right) - 2\alpha\right)\right). \tag{7.59}$$

With the values of $\psi = -0.23$ rad and $\alpha = 0.63$ rad, the relative phase $\left(\left(\psi + \frac{\pi}{2}\right) - 2\alpha\right)$ is 0.08 rad. With the relative phase smaller than $\pi/2$ and larger than $-\pi/2$, it follows from what has been stated in Section 7.5 that the velocity is flood dominant. This agrees with Fig. 7.3, in which the leading- and first-order velocity and the sum of the two are presented.

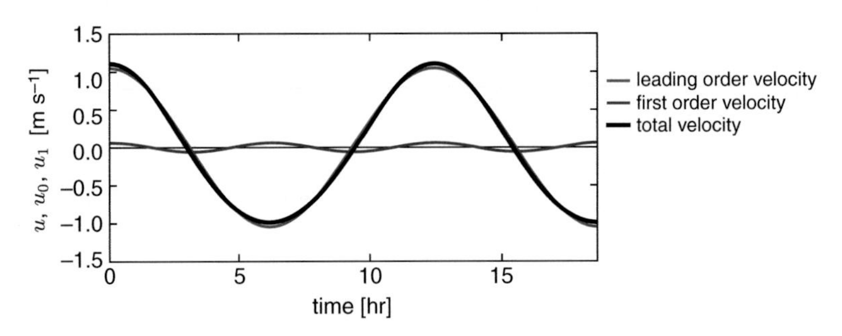

Figure 7.3 Leading-order velocity, u_0, first-order velocity, u_1, and their sum u for the representative inlet.

7.A Reduced System of Equations and Perturbation Analysis

This appendix is an extension of Section 7.2 and shows the derivation of the leading- and first-order equations. As a first step, the variables in the system of equations (7.2) and (7.6) are scaled. Similar to the scaling analysis in Section 6.3.2, the water level scale is $\hat{\eta}_0$, the velocity scale is $U = \sigma \hat{\eta}_0 A_b / bh$ and the timescale is σ^{-1}, leading to the following non-dimensional variables:

$$\eta_0^* = \frac{\eta_0}{\hat{\eta}_0}, \quad \eta_b^* = \frac{\eta_b}{\hat{\eta}_0}, \quad \eta_m^* = \frac{\eta_m}{\hat{\eta}_0}, \quad u^* = \frac{bh}{\sigma \hat{\eta}_0 A_b} u, \quad t^* = \sigma t.$$

Substituting these scaled variables into Eqs. (7.2) and (7.6) leads to the non-dimensional equations for the water motion

$$\frac{\sigma A_b}{gbh} \left[\sigma L \frac{du^*}{dt^*} + \left(\frac{m_l}{2} + \frac{F_l L}{h} \frac{1}{(1 + \epsilon \eta_m^*)} \right) u^* \right] = \eta_0^* - \eta_b^*, \tag{7.A.1}$$

and

$$\left(1 + \epsilon \eta_m^*\right) u^* = \frac{d\eta_b^*}{dt^*}. \tag{7.A.2}$$

In Eqs. (7.A.1) and (7.A.2) the dimensionless parameter ϵ denotes the ratio of the ocean tidal amplitude $\hat{\eta}_0$ and the water depth h. With $\epsilon = \hat{\eta}_0 / h \ll 1$, the following approximation can be made: $1/(1 + \epsilon \eta_m^*) = 1 - \epsilon \eta_m^* + \mathcal{O}(\epsilon^2)$. Neglecting terms of $\mathcal{O}(\epsilon^2)$ and higher, Eq. (7.A.1) reduces to

$$\frac{\sigma A_b}{gbh} \left[\sigma L \frac{du^*}{dt^*} + \left(\frac{m_l}{2} + \frac{F_l L}{h} \right) u^* - \frac{F_l L}{h} \epsilon \eta_m^* u^* \right] = \eta_0^* - \eta_b^*. \tag{7.A.3}$$

The nonlinear product $\eta_m^* u^*$ in Eqs. (7.A.2) and (7.A.3) is the result of including variation in depth with tidal stage.

The solution to Eqs. (7.A.2) and (7.A.3) is written in terms of a power series of the small parameter ϵ, i.e.

$$u^* = \epsilon^0 u_0^* + \epsilon^1 u_1^* + \mathcal{O}(\epsilon^2), \tag{7.A.4}$$

$$\eta_b^* = \epsilon^0 \eta_{b_0}^* + \epsilon^1 \eta_{b_1}^* + \mathcal{O}(\epsilon^2), \tag{7.A.5}$$

$$\eta_m^* = \epsilon^0 \eta_{m_0}^* + \epsilon^1 \eta_{m_1}^* + \mathcal{O}(\epsilon^2). \tag{7.A.6}$$

Terms with ϵ^0 are of leading-order and terms with ϵ^1 are of first-order. The scale of a first-order non-dimensional variable is ϵ times the scale of the corresponding leading-order non-dimensional variable. As a result, all non-dimensional variables in the equations are of $\mathcal{O}(1)$. As an example, the scale of the leading-order velocity, u_0, is $bh/\sigma \hat{\eta}_0 A_b$ and the scale of the first-order velocity, u_1, is $\epsilon bh/\sigma \hat{\eta}_0 A_b$. Substituting the expansions of u^*, η_b^* and η_m^*, Eqs. (7.A.4), (7.A.5) and (7.A.6) respectively, in Eqs. (7.A.2) and (7.A.3) and collecting terms of equal powers of ϵ, results in a set of equations with non-dimensional terms of $\mathcal{O}(\epsilon^0)$ and one with non-dimensional terms of $\mathcal{O}(\epsilon^1)$. Using the appropriate scales, the two sets of equations are written in terms of dimensional variables.

The resulting the system of equations at leading-order, $\mathcal{O}(\epsilon^0)$, is

$$\frac{L}{g}\frac{du_0}{dt} + F' u_0 = \eta_0 - \eta_{b_0}, \tag{7.A.7}$$

$$A_b \frac{d\eta_{b_0}}{dt} - bhu_0 = 0, \tag{7.A.8}$$

where F' is given by Eq. (7.10). The system of equations at first-order, $\mathcal{O}(\epsilon^1)$, is

$$\frac{h^2}{F_l}\frac{du_1}{dt} + \left(\frac{m_l h^2}{2F_l L} + h\right) u_1 + \frac{gh^2}{F_l L}\eta_{b_1} = u_0 \eta_{m_0}, \tag{7.A.9}$$

$$\frac{A_b}{b}\frac{d\eta_{b_1}}{dt} - hu_1 = u_0 \eta_{m_0}. \tag{7.A.10}$$

8

Cross-Sectional Stability of a Single Inlet System

8.1 Introduction

This chapter deals with the equilibrium and stability of the cross-sectional area of the inlet. The cross-sectional area is in equilibrium with the hydrodynamic environment when the volume of sand entering equals the volume of sand leaving the inlet. The equilibrium is dynamic rather than static; the inlet cross-sectional area oscillates about a mean value. The mean value is referred to as the equilibrium cross-sectional area. The equilibrium is stable when, after a perturbation, the cross-sectional area returns to its equilibrium value.

Predicting the equilibrium cross-sectional area and its stability requires coupling of hydrodynamics and sand transport. Although progress has been made, the morphodynamic models containing this element are still in a developmental state, as discussed in Chapter 11. In the meantime, recourse is taken to a simple semi-empirical approach in which the hydrodynamics is described by Eqs. (6.1) and (6.2) and the sediment dynamics is introduced empirically. This approach was first proposed by Escoffier (1940).

8.2 Equilibrium and Stability

8.2.1 Escoffier Stability Model

The Escoffier Stability Model is described in Escoffier (1940). Escoffier assumed a sinusoidal inlet velocity and reasoned that when the inlet is in equilibrium with the hydrodynamic environment, the velocity amplitude $\hat{u}$ equals the equilibrium velocity $\hat{u}_{eq}$,

$$\hat{u} = \hat{u}_{eq}. \tag{8.1}$$

In general, both $\hat{u}$ and $\hat{u}_{eq}$ are functions of the cross-sectional area, A. When the amplitude of the inlet velocity is larger than the equilibrium velocity, the inlet cross-sectional area increases and when the amplitude is smaller, the inlet cross-sectional area decreases. Escoffier took $\hat{u}_{eq} \sim 1$ m s^{-1}.

Partly based on earlier work by Sorensen (1977), the Escoffier Stability Model is expanded in van de Kreeke (2004). Because the inlet velocity for most applications is not sinusoidal, the velocity amplitude is defined in terms of the tidal prism by Eq. (5.6). The tidal prism in this equation is calculated by solving for $u(A)$ from Eqs. (6.1) and (6.2) and integrating over the ebb cycle. A rational expression for the equilibrium velocity is introduced by substituting for P/A in Eq. (5.6) from one of the A–P relationship discussed in Section 5.2. Using Eq. (5.1) as the A–P relationship the equilibrium velocity follows from Eq. (5.12). Using Eq. (5.2) as the A–P relationship, the equilibrium velocity follows from Eq. (5.13).

To calculate $u(A)$ from Eqs. (6.1) and (6.2), the hydraulic radius, R, in Eq. (6.1) is expressed in terms of the cross-sectional area, A, by using the assumption of geometric similarity (O'Brien and Dean, 1972). Referring to Appendix 8.A, geometric similarity implies that the hydraulic radius is proportional to the square root of the cross-sectional area, i.e., $R = \beta_2\sqrt{A}$, with β_2 a shape factor. As a note of caution, although there are indications that cross-sections are geometrically similar (Winton and Mehta, 1981), it seems wise to heed the observation of (Walton, 2004a) that the assumption of geometric similarity is not trivial and to also calculate the closure curve for hydraulic radius cross-sectional area scenarios other than geometric similarity.

8.2.2 Escoffier Diagram

The Escoffier Diagram consists of a closure curve, $\hat{u}(A)$ and an equilibrium velocity curve, $\hat{u}_{eq}(A)$. The closure curve represents the relationship between velocity amplitude and cross-sectional area. The equilibrium velocity curve represents the relationship between equilibrium velocity and cross-sectional area. The diagram is used to calculate the equilibrium cross-sectional area(s) and their stability. As an example, the Escoffier Diagram for the representative inlet is presented in Fig. 8.1. The parameter values for the representative inlet are given in Table 6.2. The closure curve is calculated using the Öszoy–Mehta Solution (Section 6.4). Assuming the A–P relationship for the representative is of the form given by Eq. (5.2), the equilibrium velocity curve is calculated using Eq. (5.13). Taking $C_l = 6.8 \times 10^{-5}$ m^{-1}, the value for the Dutch Wadden Sea inlets, and the tidal period $T = 44,712$ s, it follows from Eq. (5.13) that $\hat{u}_{eq} = 1.03$ m s^{-1}. The equilibrium velocity is independent of the cross-sectional area. Selecting Eq. (5.1) rather than Eq. (5.2) as the A–P relationship, the equilibrium velocity is given by Eq. (5.12). This makes $\hat{u}_{eq}$ dependent on the cross-sectional area. The intersections of the closure curve and equilibrium velocity curve correspond to equilibrium cross-sectional areas. For equilibrium cross-sectional areas to exist, the two curves must intersect. In

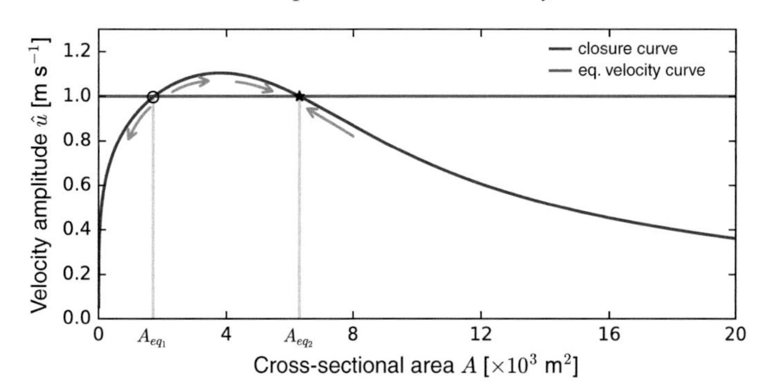

Figure 8.1 Escoffier Diagram for the representative inlet; for parameter values reference is made to Table 6.2.

that case there are always two equilibrium cross-sectional areas, $A_{1_{eq}}$ and $A_{2_{eq}}$. In the example for the representative inlet, $A_{1_{eq}} = 1,700$ m^2 and $A_{2_{eq}} = 6,300$ m^2.

In calculating the closure curve and equilibrium velocity curve, the velocity amplitude, $\hat{u}(A)$, and the A–P relationship used to calculate the equilibrium velocity, $\hat{u}_{eq}(A)$, must pertain to the same (mean or spring) tide conditions.

Using the Escoffier Diagram, the stability of the equilibriums is evaluated realizing that for $\hat{u} > \hat{u}_{eq}$ the inlet erodes and for $\hat{u} < \hat{u}_{eq}$ the inlet shoals (Section 5.2.2). It then follows that the equilibrium corresponding to A_{eq_1} is unstable; a decrease in cross-sectional area leads to closure and an increase causes the cross-sectional area to increase until it reaches the value of A_{eq_2}. The equilibrium corresponding to A_{eq_2} is stable; after a decrease in cross-sectional area, and provided its value remains larger than A_{eq_1}, the cross-sectional area returns to the equilibrium value, when increasing the cross-sectional area it will always return to the equilibrium value. The arrows in Fig. 8.1 show the direction in which the cross-sectional areas evolve after a perturbation. It follows from the Escoffier Diagram that the smaller the equilibrium velocity, the larger the stable equilibrium cross-sectional area. The difference between the stable and unstable equilibrium cross-sectional areas may be looked upon as a measure for the stability of the cross-sectional area A_{eq_2}. As shown in Appendix 8.B, the stability of the two equilibrium cross-sectional areas can be more formally investigated using a linear stability analysis.

The Escoffier Diagram is not only a valuable tool in determining the equilibrium cross-sectional areas and their stability, but it also provides insight into the response of the tidal inlet when the equilibrium is disturbed.

8.2.3 The Shape of the Closure Curve

The shape of the closure curve in Fig. 8.1 is typical for most tidal inlets. This is explained by using approximate expressions for the velocity amplitude for small and large values of the cross-sectional area. For small values of the cross-sectional area, the friction term in Eq. (6.1) is large compared to the inertia term. Neglecting the inertia term, the expression for the dimensionless velocity amplitude $\hat{u}^*$ is given by Eq. (6.27). Furthermore, for small values of A the repletion coefficient K_1, given by Eq. (6.9), approaches zero. Neglecting entrance and exit losses, replacing the hydraulic radius by $R = \beta_2\sqrt{A}$, it then follows that the dimensional velocity amplitude is

$$\hat{u} \cong \sqrt{\frac{3\pi g\beta_2\sqrt{A}}{8FL}\hat{\eta}_0}. \tag{8.2}$$

The velocity amplitude $\hat{u}$ approaches zero when A goes to zero and increases as $A^{1/4}$ for increasing values of A.

For large values of A, the basin water level approaches the ocean water level and the tidal prism attains a value

$$P = 2A_b\hat{\eta}_0. \tag{8.3}$$

Using Eqs. (5.6) and (8.3), the velocity amplitude for large values of the cross-sectional area is

$$\hat{u} = \frac{2\pi A_b\hat{\eta}_0}{AT}. \tag{8.4}$$

The velocity amplitude decreases as A^{-1} for increasing values of A.

8.3 Adaptation Timescale

The adaptation timescale is a measure for the time it takes the cross-sectional area to return to its original equilibrium after a perturbation. To determine the adaptation timescale, it is assumed that, as a result of a storm, an excess volume of sand is deposited in the entrance section of the inlet, thereby reducing the cross-sectional area. The entrance section has a length L_e. Following the storm, the excess volume of sand and the fraction M' of the gross longshore sand transport that continues to enter the inlet are gradually removed by the ebb tidal current until the cross-sectional area again attains its equilibrium value. Taking the sand transport by the ebb tidal current proportional to the velocity amplitude to the power n, the conservation of sand equation for the entrance section is

$$L_e\frac{dA}{dt} = k\hat{u}^n - M'. \tag{8.5}$$

When at equilibrium, this equation reduces to

$$k\hat{u}_{eq}^n = M'. \tag{8.6}$$

Eliminating k between Eqs. (8.5) and (8.6), it follows that

$$\frac{dA}{dt} = \frac{M'}{L_e}\left[\left(\frac{\hat{u}}{\hat{u}_{eq}}\right)^n - 1\right]. \tag{8.7}$$

In this equation the velocity amplitude $\hat{u}$ is a function of the cross-sectional area A. This makes solving Eq. (8.7) a somewhat cumbersome (numerical) exercise.

To demonstrate the nature of the solution, the simplifying assumption is made that during the adaptation period the tidal prism remains constant. Using Eq. (5.6), this implies

$$\frac{\hat{u}}{\hat{u}_{eq}} = \frac{A_{eq}}{A}. \tag{8.8}$$

Substituting for $\hat{u}/\hat{u}_{eq}$ in Eq. (8.7) results in

$$\frac{dA}{dt} = \frac{M'}{L_e}\left[\left(\frac{A_{eq}}{A}\right)^n - 1\right]. \tag{8.9}$$

The quotient $1/A^n$ is linearized by introducing

$$A = A_{eq} + \Delta A, \tag{8.10}$$

and thus

$$\frac{1}{A^n} = \frac{1}{A_{eq}^n\left(1 + \frac{\Delta A}{A_{eq}}\right)^n}. \tag{8.11}$$

With

$$\frac{\Delta A}{A_{eq}} \ll 1, \tag{8.12}$$

Eq. (8.11) is approximated by

$$\frac{1}{A^n} \cong \frac{1}{A_{eq}^n}\left(1 - n\left(\frac{\Delta A}{A_{eq}}\right)\right). \tag{8.13}$$

Using Eq. (8.10), it follows from Eq. (8.13) that

$$\frac{A_{eq}^n}{A^n} \cong \left(1 + n\left(\frac{A_{eq} - A}{A_{eq}}\right)\right). \tag{8.14}$$

Substituting for A_{eq}^n/A^n from Eq. (8.14) in Eq. (8.9) results in

$$\frac{d(A - A_{eq})}{dt} + \frac{nM'}{L_e A_{eq}}(A - A_{eq}) = 0. \tag{8.15}$$

From Eq. (8.15) it follows that the cross-sectional area adjusts exponentially. The time for the initial deviation to reduce by a factor e (the Naperian timescale) is

$$\tau = \frac{L_e A_{eq}}{nM'}. \tag{8.16}$$

For the same transport of sand, M', entering the inlet, the adaptation timescale decreases with decreasing values of cross-sectional area; smaller inlets adapt faster than larger inlets.

As an example, for the Frisian Inlet, one of the inlets of the Dutch Wadden Sea, with $A_{eq} = 15,300$ m^2 (van de Kreeke, 2004), a storm deposit of 50,000 m^3 extending over a distance $L_e = 50$ m reduces the cross-sectional area by about 6.5 percent. With $M' = 500,000$ m^3 year^{-1} and n between 3 and 5, it then follows from Eq. (8.16) that the timescale for the cross-sectional area to return to its equilibrium value is 0.3–0.5 year. Similar values are reported in Kraus (1998).

In the foregoing, it is assumed that, as a result of a storm, a volume of sand is deposited in the entrance section of an inlet. The same reasoning and results hold when assuming that, as a result of a storm, a volume of sand is removed from the entrance section of the inlet.

8.4 Cross-Sectional Stability of Pass Cavallo

An aerial view of Pass Cavallo is presented in Fig. 8.2. Until 1963 Pass Cavallo was the sole inlet connecting Matagorda Bay (TX) and the Gulf of Mexico. In 1963 a second inlet, the Matagorda shipping channel, was artificially opened. This affected the stability of Pass Cavallo. Here the equilibrium and stability of Pass Cavallo prior to opening of the second inlet is evaluated using the Escoffier Stability Model described in Section 8.2.1. The analysis is in part based on an earlier study by van de Kreeke (1985), with additional data derived from later studies by Kraus et al. (2006) and Batten et al. (2007).

Prior to the opening of the second inlet, Pass Cavallo was in stable equilibrium with the hydrodynamic environment. It had been open for at least 200 years. Using observations from 1959, the cross-sectional area was 8,000 m^2. The cross-section had a parabolic shape with a width to maximum depth ratio of 333. Offshore tides are dominantly diurnal with a great diurnal amplitude of 0.27 m. The great diurnal amplitude is half the distance between the tidal datum planes of Mean Higher High (MHH) and Mean Lower Low (MLL). It is used to characterize the tide in areas where it has a mixed character (Jarrett, 1976). The observed tidal prism corresponding to the great diurnal tide is $P = 2.1 \times 10^8$ m^3. Compared to the tidal prism, freshwater inflow is negligible. The observed maximum velocity corresponding to the great diurnal tide is 1 m s^{-1}.

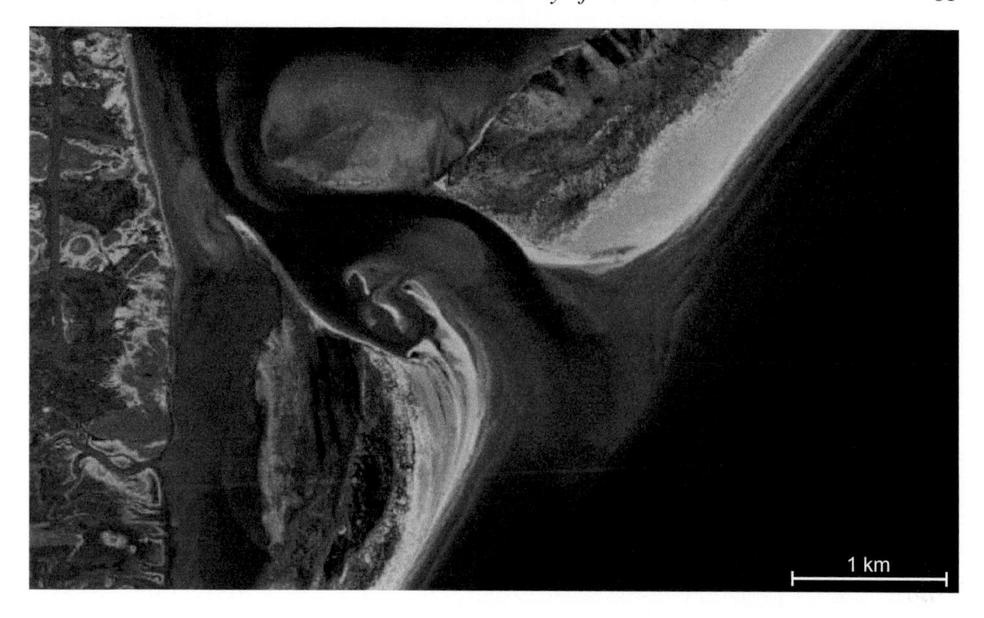

Figure 8.2 Pass Cavallo (TX) in 2015 (Esri et al., 2016).

Lacking an A–P relationship, the equilibrium velocity for Pass Cavallo is determined from the observed values of the tidal prism $P = 2.1 \times 10^8$ m^3 and the cross-sectional area $A = 8,000$ m^2 at the time the inlet was at equilibrium. Using Eq. (5.6) with $T = 86,400$ s results in a value of the equilibrium velocity $\hat{u}_{eq} = 0.96$ m s^{-1}. The closure curve $\hat{u}(A)$ is calculated using Eq. (5.6) with the parameter values given in Table 8.1. The value of P in Eq. (5.6) is calculated from the Keulegan Solution (Section 6.5), realizing that $P = 2\hat{\eta}_0 A_b \sin\tau$, with $\sin\tau$ a function of the repletion coefficient K_1 tabulated in Table 6.1. In calculating the values of the repletion coefficient for different values of the cross-sectional areas, cross-sections are assumed to be geometrically similar with the value of the shape factor given in Table 8.1 (Appendix 8.A). With a relatively large basin surface area of 9.4×10^8 m^2 and shallow depth, the assumption of a pumping mode for Matagorda Bay, as used in the Keulegan equations, is questionable. To account for this and to match the observed and calculated tidal prism, the basin surface area is reduced to a value of 4.0×10^8 m^2. In this respect the basin surface area serves as a calibration parameter.

The Escoffier Diagram for Pass Cavallo is presented in Fig. 8.3. The value of the stable equilibrium cross-sectional area agrees with the observed cross-sectional area of 8,000 m^2. The value of the unstable equilibrium is approximately 300 m^2. The large difference between the values of the stable and unstable equilibrium cross-sectional areas explains why the inlet has remained open for at least 200 years despite several tropical storms and hurricanes.

Table 8.1 *Parameter values for Pass Cavallo prior to opening of the Matagorda shipping channel.*

Parameter	Symbol	Dimension	Value
Inlet length	L	m	1,600
Friction factor	F	–	2.5×10^{-3}
Entrance/exit loss coefficient	m	–	0.5
Shape factor	β_2	–	0.045
Basin surface area	A_b	m^2	4×10^8
Tidal amplitude	$\hat{\eta}_0$	m	0.27
Tidal frequency	σ	rad s^{-1}	7.3×10^{-5}

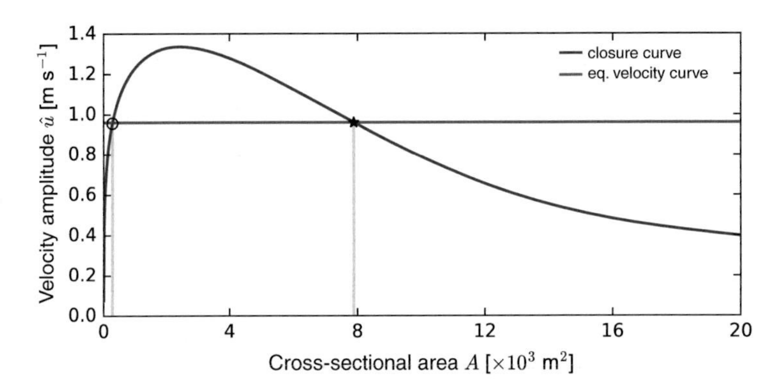

Figure 8.3 Escoffier Diagram for Pass Cavallo. For parameter values reference is made to Table 8.1.

As for many inlets, parameter values used in constructing the Escoffier Diagram for Pass Cavallo are at best estimates. In spite of this, at a minimum the diagram explains the observed equilibrium cross-sectional area and its stability.

8.A Geometric Similarity

Objects that are geometrically similar have the same shape, and the ratio of any two linear dimensions is the same. An example is the triangles in Fig. 8.A.1. A property of geometrically similar objects is that any linear dimension is proportional to the square root of the area. This relationship is demonstrated for the widths and hydraulic radii of the triangular cross-sections.

The two triangles in the figure have the same shape. The shape is determined by the two ratios

$$\frac{L_1}{h} = \alpha_1, \qquad \text{and} \qquad \frac{L_2}{h} = \alpha_2. \tag{8.A.1}$$

The ratios α_1 and α_2 for the larger cross-section are the same as the corresponding ratios for the smaller cross-section.

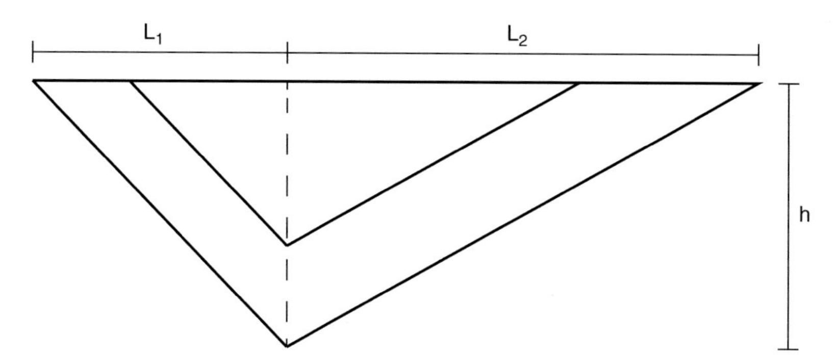

Figure 8.A.1 Geometrically similar triangles.

Starting with the width, $W = L_1 + L_2$, of the cross-section

$$\frac{W}{h} = \alpha_1 + \alpha_2. \tag{8.A.2}$$

The cross-sectional area is

$$A = \tfrac{1}{2} h W. \tag{8.A.3}$$

Eliminating h between Eqs. (8.A.2) and (8.A.3) results in

$$W = \beta_1 \sqrt{A}, \tag{8.A.4}$$

with

$$\beta_1 = \sqrt{2(\alpha_1 + \alpha_2)}. \tag{8.A.5}$$

The hydraulic radius R is the ratio of the cross-sectional area A and the wetted perimeter P,

$$R = \frac{A}{P}. \tag{8.A.6}$$

In terms of the ratios α_1 and α_2 the cross-sectional area is

$$A = \tfrac{1}{2}(\alpha_1 + \alpha_2) h^2, \tag{8.A.7}$$

and the wetted perimeter is

$$P = h\left(\sqrt{\alpha_1^2 + 1} + \sqrt{\alpha_2^2 + 1}\right). \tag{8.A.8}$$

The expression for the hydraulic radius then is

$$R = \frac{\tfrac{1}{2} h (\alpha_1 + \alpha_2)}{\left(\sqrt{\alpha_1^2 + 1} + \sqrt{\alpha_2^2 + 1}\right)}. \tag{8.A.9}$$

Eliminating W between Eqs. (8.A.2) and (8.A.3) results in

$$h = \sqrt{\frac{2}{(\alpha_1 + \alpha_2)}}\sqrt{A}. \tag{8.A.10}$$

Substituting for h from Eq. (8.A.10) in Eq. (8.A.9), it follows that

$$R = \beta_2 \sqrt{A}, \tag{8.A.11}$$

with

$$\beta_2 = \frac{\frac{1}{\sqrt{2}}\sqrt{\alpha_1 + \alpha_2}}{\left(\sqrt{\alpha_1^2 + 1} + \sqrt{\alpha_2^2 + 1}\right)}. \tag{8.A.12}$$

The coefficients β_1 and β_2 are referred to as *shape factors*.

8.B Linear Stability Analysis

When removed from equilibrium, the rate of change of cross-sectional area is given by Eq. (8.7) in the main text, i.e.,

$$\frac{dA}{dt} = \frac{M'}{L_e}\left[\left(\frac{\hat{u}}{\hat{u}_{eq}}\right)^n - 1\right]. \tag{8.B.1}$$

The right-hand side of this equation is linearized by defining

$$\hat{u} = \hat{u}_{eq} + \Delta\hat{u}. \tag{8.B.2}$$

From Eq. (8.B.2) it follows that

$$\left(\frac{\hat{u}}{\hat{u}_{eq}}\right)^n = \left(1 + \frac{\Delta\hat{u}}{\hat{u}_{eq}}\right)^n. \tag{8.B.3}$$

For small deviations from equilibrium, i.e., $\Delta\hat{u}/\hat{u}_{eq} \ll 1$,

$$\left(1 + \frac{\Delta\hat{u}}{\hat{u}_{eq}}\right)^n \cong 1 - n + n\frac{\hat{u}}{\hat{u}_{eq}}, \tag{8.B.4}$$

and thus

$$\left(\frac{\hat{u}}{\hat{u}_{eq}}\right)^n - 1 \cong \frac{n(\hat{u} - \hat{u}_{eq})}{\hat{u}_{eq}}. \tag{8.B.5}$$

Furthermore,

$$(\hat{u} - \hat{u}_{eq}) \cong \frac{\partial\hat{u}}{\partial A}(A - A_{eq}), \tag{8.B.6}$$

where $\partial \hat{u}/\partial A$ is evaluated at the equilibrium, $A = A_{eq}$. Using Eqs. (8.B.5) and (8.B.6), Eq. (8.B.1) is written as

$$\frac{d(A - A_{eq})}{dt} = \lambda(A - A_{eq}), \tag{8.B.7}$$

with

$$\lambda = \frac{nM}{\hat{u}_{eq}L}\frac{\partial \hat{u}}{\partial A}. \tag{8.B.8}$$

From Eq. (8.B.7)

$$(A - A_{eq}) = (A - A_{eq})_0 e^{\lambda t}. \tag{8.B.9}$$

In Eq. (8.B.9), $(A - A_{eq})_0$ is the initial perturbation. The perturbation increases for λ is positive and decreases for λ is negative. It follows from Eq. (8.B.8) that the sign of λ is determined by the sign of $\partial \hat{u}/\partial A$, evaluated at $A = A_{eq}$. Referring to the Escoffier Diagram, Fig. 8.1 in the main text, it follows that with $\partial \hat{u}/\partial A$ positive, the equilibrium $A = A_{eq_1}$ is unstable and with $\partial \hat{u}/\partial A$ negative, the equilibrium $A = A_{eq_2}$ is stable.

9

Cross-Sectional Stability of a Double Inlet System, Assuming a Uniformly Varying Basin Water Level

9.1 Introduction

Instead of one inlet, many back-barrier lagoons, bays and inland seas are connected to the ocean by multiple inlets. Examples are the Ría Formosa in south Portugal (Salles et al., 2005), the Dutch, German and Danish Wadden Sea (Ehlers, 1988) and the Venice Lagoon in Italy (Tambroni and Seminara, 2006); see Fig. 1.1. This chapter concerns the interaction of these inlets, with emphasis on cross-sectional stability.

Depending on the hydraulic efficiency, inlets connecting the same back-barrier lagoon capture a smaller or larger part of the tidal prism. The tidal prism is the volume of water entering and leaving the back-barrier lagoon during a tidal cycle. The fraction of the tidal prism entering and leaving an individual inlet is the prime parameter determining the cross-sectional stability of that inlet. If this fraction is too small, the inlet closes. In this respect the opening of a new inlet is of interest. Potentially, the new inlet could lead to a decrease of the tidal prisms of the already existing inlet(s) and chances are that some of these inlets close. In that case, it has to be decided to either close the new inlet or leave it open.

An example of a recently opened inlet is the Breach at Old Inlet on Fire Island, NY (Fig. 2.1). The inlet was opened in October 2012 during hurricane Sandy. Together with the already existing Fire Island Inlet, the Breach at Old Inlet connects Great South Bay to the ocean. To determine the behavior of the new inlet and its effect on Fire Island Inlet, both inlets are being monitored (National Park Service, 2012). As of March 2015, the Breach at Old Inlet was still open. Based on the monitoring results it might be possible to arrive at a rational decision to either close the inlet or leave it open.

In this chapter, as a first step to determine the cross-sectional stability of tidal inlets connecting the same back-barrier lagoon to the ocean, the method used to determine the cross-sectional stability of a single inlet system (Chapter 8) is expanded and applied to a double inlet system.

9.2 Escoffier Stability Model for a Double Inlet System

The method to evaluate the cross-sectional stability of a double inlet system is an extension of the Escoffier Stability Model for a single inlet presented in Section 8.2.1. In the following, the various steps in the stability analysis are described using the Texel-Vlie inlet system as an example (Brouwer, 2013; Brouwer et al., 2008; van de Kreeke et al., 2008). This double inlet system is part of the Dutch Wadden Sea and is shown in Fig. 9.1.

9.2.1 Schematization

For the purposes of illustrating the stability analysis, the Texel-Vlie inlet system is schematized to a basin with two inlets (Fig. 9.2). The schematized system is symmetric with parameter values presented in Table 9.1. Similar to the single inlet, inlets are prismatic with diverging parts on both ends (Fig. 6.A.1).

Parameter values are the same for both inlets. Assumed is a triangular cross-section with a surface width to maximum depth ratio of 115, resulting in a shape

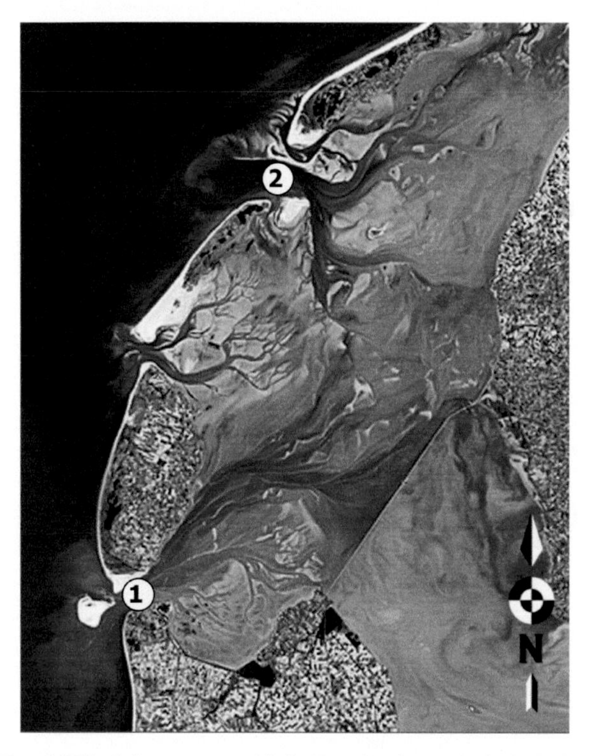

Figure 9.1 Texel-Vlie inlet system; (1) is Texel Inlet and (2) is Vlie Inlet (USGS and ESA, 2011).

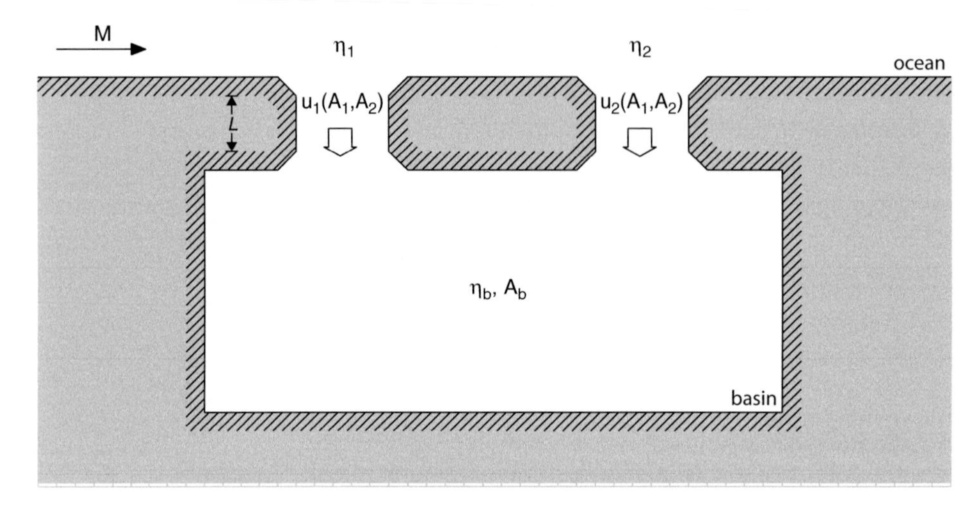

Figure 9.2 Schematization of Texel-Vlie inlet system.

Table 9.1 *Parameter values for the schematized Texel-Vlie inlet system.*

Parameter	Symbol	Dimension	Value
Inlet length	L	m	5,000
Friction factor	F	[-]	4×10^{-3}
Entrance/exit loss coefficient	m	[-]	0
Shape factor	β_2	[-]	0.065
Basin surface area	A_b	m^2	14×10^8
Tidal amplitude	$\hat{\eta}_0$	m	0.8
Tidal frequency	σ	rad s^{-1}	1.4×10^{-4}
Equilibrium velocity	$\hat{u}_{eq}$	m s^{-1}	1.0

factor $\beta_2 = 0.065$ (Appendix 8.A). The ocean tidal amplitude pertains to mean tide conditions.

9.2.2 Equilibrium Velocity

Similar to Eq. (8.1) for a single inlet system, the condition for equilibrium of a double inlet system is

$$\hat{u}_j(A_1, A_2) = \hat{u}_{eq}, \qquad j = 1, 2. \tag{9.1}$$

In this equation $j = 1$ refers to the Texel Inlet and $j = 2$ refers to the Vlie Inlet. The velocity amplitude $\hat{u}_j$ is a function of the cross-sectional areas, A, of both inlets. In analogy with the closure curve $\hat{u}(A)$ for the single inlet system (Fig. 8.1), the surfaces $\hat{u}_j(A_1, A_2)$ are referred to as closure surfaces. The velocity $\hat{u}_{eq}$ is the equilibrium velocity defined in Section 5.2.4. Depending on the cross-sectional area – tidal prism relationship, values of $\hat{u}_{eq}$ follow from Eqs. (5.12) or (5.13). Taking Eq. (5.13) as an example, the equilibrium velocity is

$$\hat{u}_{eq} = \frac{\pi}{C_l T}, \tag{9.2}$$

where T is tidal period and C_l is the coefficient in the cross-sectional area – tidal prism relationship given by Eq. (5.2). For the inlets of the Dutch Wadden Sea with P in Eq. (5.2) referring to mean tide conditions, $C_l = 6.8 \times 10^{-5}$ m^{-1} (Section 5.2.3). From Eq. (9.2), with $T = 44,712$ s, it follows that $\hat{u}_{eq} = 1.03$ m s^{-1}.

9.2.3 Governing Equations

The momentum equation for the individual inlets is Eq. (6.1). Assuming evolving cross-sections remain geometrically similar, the hydraulic radius, R, in this equation is taken proportional to the square root of the cross-sectional area resulting in

$$\frac{L_j}{g}\frac{du_j}{dt} + \left(\frac{m}{2g} + \frac{FL}{g\beta_2\sqrt{A_j}}\right) u_j|u_j| = \eta_0 - \eta_b, \qquad j = 1, 2. \tag{9.3}$$

Assuming a uniformly fluctuating basin water level, the continuity equation is

$$A_1 u_1 + A_2 u_2 = A_b \frac{d\eta_b}{dt}. \tag{9.4}$$

In these equations, L is the length of the inlet, g is gravity acceleration, u is the cross-sectionally averaged velocity, t is time, m is an entrance/exit loss coefficient, F is a bottom friction factor, β_2 is a shape factor (Appendix 8.A), A_b is the basin surface area, η_0 is the ocean tide and η_b is the basin tide. In Eqs. (9.3) and (9.4), variations in cross-sectional area with tidal stage are neglected. The ocean tide is simple harmonic with the same amplitude and phase for both inlets, i.e.,

$$\eta_0 = \hat{\eta}_0 \sin \sigma t. \tag{9.5}$$

Using the terminology introduced in Chapter 1, the system of equations constitutes a process-based exploratory model.

9.2.4 Closure Surface

Given the cross-sectional areas and the parameter values for the schematized Texel-Vlie inlet system in Table 9.1, the governing equations are solved for the basin water level and inlet velocities. For this a finite difference solution can be used as described in (Brouwer, 2006). In the present application a more efficient semi-analytical solution, described in Appendix 2A of (Brouwer, 2013), is used. The semi-analytical solution leads to inlet velocities that are sinusoidal.

Using the results of the semi-analytical solution, the closure surfaces, $\hat{u}_1(A_1, A_2)$ for Inlet 1 and $\hat{u}_2(A_1, A_2)$ for Inlet 2 are presented in, respectively, Figs. 9.3a and 9.3b. Referring to Fig. 9.3a, for a constant value of A_2, the variation in $\hat{u}_1$ with A_1 resembles the closure curve in the Escoffier Diagram presented in Fig. 8.1. For a constant value of A_1, $\hat{u}_1$ monotonically decreases with increasing values of A_2. Similarly referring to Fig. 9.3b, keeping A_1 constant, the variation in $\hat{u}_2$ with A_2 resembles the closure curve in the Escoffier Diagram. For constant values of A_2, $\hat{u}_2$ monotonically decreases with increasing values of A_1. The solid black line in Fig. 9.3a is the intersection of the plane $\hat{u}_1 = \hat{u}_{eq}$ with the closure surface of Inlet 1. Similarly, the solid black line in Fig. 9.3b is the intersection of the plane $\hat{u}_2 = \hat{u}_{eq}$ and the closure surface of Inlet 2.

9.2.5 Equilibrium Velocity Curves

The equilibrium velocity curve for Inlet 1 is the projection of the solid black line in Fig. 9.3a in the (A_1, A_2)-plane. It is the locus of (A_1, A_2)-values for

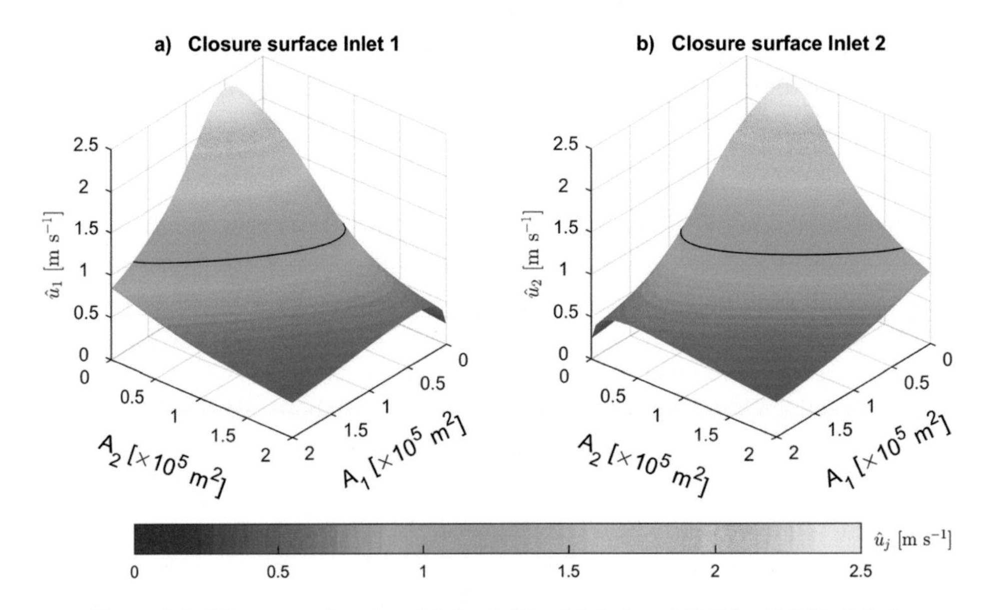

Figure 9.3 Closure surface for a) Inlet 1 (Texel Inlet) and b) Inlet 2 (Vlie Inlet).

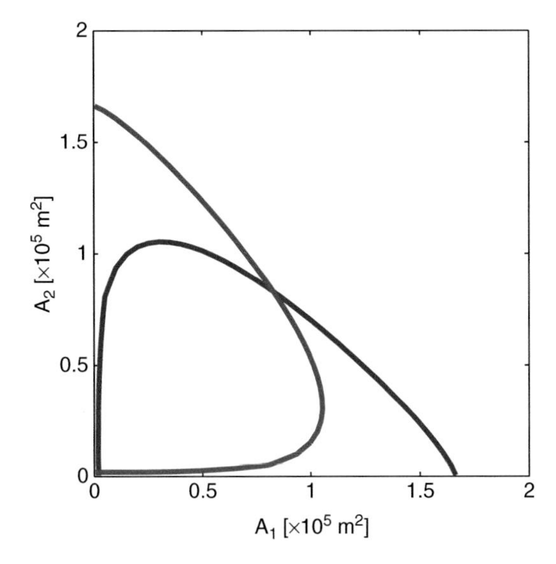

Figure 9.4 Equilibrium velocity curves for Texel Inlet (blue) and Vlie Inlet (red). (From: Brouwer, 2013)

which $\hat{u}_1 = \hat{u}_{eq}$. Similarly, projecting the solid black line in Fig. 9.3b in the (A_1, A_2)-plane results in the equilibrium velocity curve for Inlet 2, representing the locus of (A_1, A_2)-values for which $\hat{u}_2 = \hat{u}_{eq}$. Both equilibrium velocity curves are presented in Fig. 9.4. Because the double inlet system is symmetric, the equilibrium velocity curves for the two inlets are symmetric with respect to the line $A_1 = A_2$. The two intersections of the equilibrium velocity curves represent combinations of (A_1, A_2) for which the velocity amplitudes in both inlets equal the equilibrium velocity and both inlets are in equilibrium with the hydrodynamic environment. The two sets of equilibrium cross-sectional areas are $(A_1 A_2) \cong (600 \text{ m}^2, 600 \text{ m}^2)$ and $(A_1 A_2) \cong (8,100 \text{ m}^2, 8,100 \text{ m}^2)$.

9.2.6 Flow Diagram

The flow diagram for the double inlet system is the counterpart of the Escoffier Diagram for the single inlet system. A flow diagram consists of the two equilibrium velocity curves together with a vector plot. The vectors represent the rate of change of the cross-sectional areas after a perturbation and are defined as

$$\frac{d\vec{A}}{dt} = \frac{dA_1}{dt}\vec{e}_1 + \frac{dA_2}{dt}\vec{e}_2, \tag{9.6}$$

where $\vec{e}_1$ and $\vec{e}_2$ are the unit vectors in, respectively, the A_1 and A_2 direction. Referring to Eq. (8.7), the rate of change of the cross-sectional areas is related to the ratio of the velocity amplitudes $\hat{u}_j$ and the equilibrium velocity $\hat{u}_{eq}$ by

$$\frac{dA_j}{dt} = \frac{M'}{L_{e_j}} \left[\left(\frac{\hat{u}_j}{\hat{u}_{eq}} \right)^n - 1 \right], \qquad j = 1, 2, \tag{9.7}$$

where L_{e_j} is the length of the entrance section of Inlet j. The entrance section is the seaward part of the inlet where, during a storm event, an excess volume of sand is deposited. M' is the fraction of the long-term average longshore sand transport entering the inlet on the flood and leaving on the ebb. The direction of the vectors in the flow diagram shows whether, after a perturbation, the system returns to the original equilibrium or moves away from it. The direction of the vectors thus determines whether the equilibrium is stable or unstable. Because it is only the direction and not the magnitude that is of interest, vectors are given a unit length. In the present example values of M' and L_{e_j} are the same for both inlets. Therefore, the ratio M'/L_{e_i} does not play a role in the direction of the vectors.

The flow diagram of the schematized Texel-Vlie inlet system, presented in Fig. 9.5, shows two equilibriums. Details in the vicinity of the two equilibriums are presented in Figs. 9.6a and 9.6b. For the equilibrium in Fig. 9.6a holds that, regardless of the direction of the perturbation, the cross-sectional areas move away from the original equilibrium; the equilibrium is unstable. For the equilibrium in Fig. 9.6b, a distinction is made between perturbations whereby both cross-sectional areas increase or decrease and perturbations whereby one cross-sectional

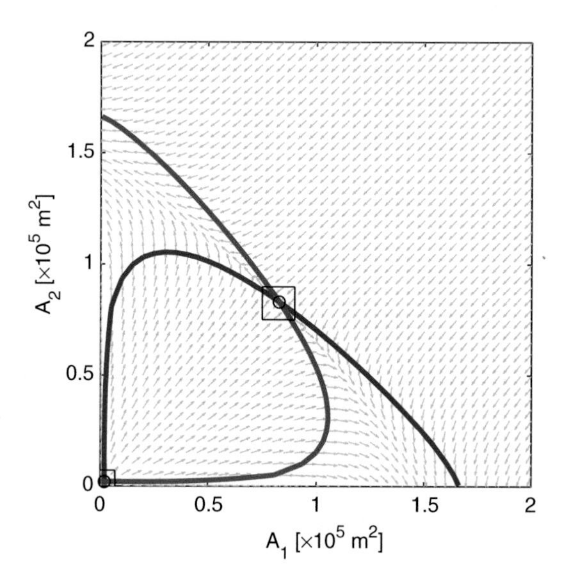

Figure 9.5 Flow diagram for the schematized Texel-Vlie inlet system. Blue is the equilibrium velocity curve of the Texel Inlet and red is the equilibrium velocity curve of the Vlie Inlet. Circles denote a set of equilibrium cross-sectional areas.

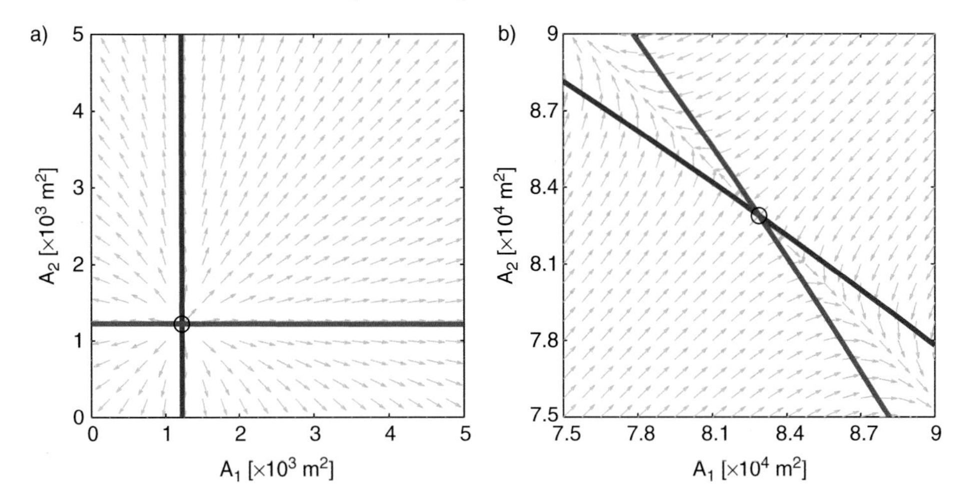

Figure 9.6 Zoomed-in flow diagram for the Texel-Vlie inlet system around equilibriums a) $(A_1, A_2) \cong (600\,\mathrm{m}^2, 600\,\mathrm{m}^2)$ and b) $(A_1, A_2) \cong (8{,}100\,\mathrm{m}^2, 8{,}100\,\mathrm{m}^2)$. Blue is the equilibrium velocity curve of the Texel Inlet and red is the equilibrium velocity curve of the Vlie Inlet.

area increases and the other decreases. When both cross-sectional areas increase or decrease, they return to the original equilibrium. When the perturbations are in opposite direction, cross-sectional areas move further away from the original equilibrium. Defining a stable equilibrium as one where, regardless of the direction of the perturbation, the cross-sectional areas return to the original equilibrium, the equilibrium in Fig. 9.6b is unstable.

In summary, the Escoffier Stability Model, with the hydrodynamics described by Eqs. (9.3) and (9.4) and the same sinusoidal ocean tide at both inlets given by Eq. (9.5), does not lead to a set of stable equilibrium areas for the Texel-Vlie inlet system.

9.3 Conditions for a Set of Stable Cross-Sectional Areas

In addition to the Texel-Vlie inlet system, the Escoffier Stability Model described in the preceding sections has been applied to the Pass Cavallo-Matagorda inlet system in Texas (van de Kreeke, 1985) and the Big Marco-Capri inlet system in Florida (van de Kreeke, 1990b). Similar to the Texel-Vlie inlet system, the equilibrium velocity curves of these double inlet systems show two intersections, representing unstable equilibriums. The absence of stable equilibriums for the Texel-Vlie, Pass Cavallo-Matagorda and Big Marco-Capri inlet systems when applying the Escoffier Stability Model is somewhat surprising, given the fact that these double inlet systems have existed for long periods of time and are considered stable.

The lack of stable equilibriums when using the Escoffier Stability Model for a double inlet system was first investigated in van de Kreeke (1990a). Similar to the preceding examples, he considered a symmetric, friction-dominated double inlet system, forced with equal ocean tides at both inlets. Based on the results of this study, he postulated four possible configurations for the two equilibrium velocity curves. Three of the configurations showed two intersections of the equilibrium velocity curves, none of them representing stable equilibriums. The fourth configuration showed four intersections, three of them representing unstable equilibriums and the fourth representing a stable equilibrium. From this it was concluded that for a stable equilibrium to exist, the equilibrium velocity curves must have four intersections. To investigate if Eqs. (9.3) and (9.4) with the boundary condition, Eq. (9.5), lead to four intersections, Eq. (9.3) was simplified by neglecting the inertia and entrance/exit loss term. As shown in Section 6.3.1, these terms are usually small compared to the bottom friction term. Using an analytical solution it was then shown that the reduced set of equations at best leads to two intersections, excluding the presence of a stable equilibrium.

Using the Escoffier Stability Model, with Eqs. (9.3) and (9.4) describing the hydrodynamics but with ocean tides that have different amplitudes for the two inlets, Brouwer et al. (2012) investigated the cross-sectional stability of the Faro-Armona double inlet system in Portugal. The equilibrium velocity curves showed four intersections, representing three unstable and one stable equilibrium. The Faro and Armona Inlets are somewhat exceptional in that the inlets are relatively short and deep. As a result, the hydrodynamic balance is between pressure gradient and entrance/exit losses rather than pressure gradient and bottom friction. How this affects the equilibrium and stability of the inlets is further discussed in Brouwer et al. (2012).

A relatively simple explanation why the Escoffier Stability Model does not lead to stable cross-sectional areas is given in Roos et al. (2013). The Escoffier Stability Model includes the assumption of a uniformly fluctuating basin level and the same ocean tide at both inlets. This implies that at all times the pressure gradients over the two inlets are the same. As shown in Roos et al. (2013) this condition prevents stable equilibrium cross-sectional areas. To explain this, they distinguish between a destabilizing and stabilizing mechanism. The first one is associated with the bottom friction in the inlets and the second one with the pressure gradients over the inlets. The way these mechanisms operate is demonstrated for a perturbation whereby the cross-sectional area of Inlet 1 is increased and the cross-sectional area of Inlet 2 is decreased. Increasing the cross-sectional area of Inlet 1 leads to a decrease in bottom friction in Inlet 1 and decreasing the cross-sectional area of Inlet 2 leads to an increase in bottom friction in Inlet 2 (see Eq. (9.3)). To explain the destabilizing mechanism, it is first assumed that after the perturbation the pressure gradients over the inlets remain the same. Assuming a friction-dominated system, the decrease in bottom friction leads to an increase in velocity and erosion in Inlet 1. Similarly, the increase in bottom friction leads to a decrease in velocity and accretion in Inlet 2. As a result, both inlets move away from equilibrium. To explain a possible return to equilibrium, the assumption of the pressure gradients over the inlets remaining the same after the perturbation is removed. For Inlet 1 to return to equilibrium, the pressure gradient needs to decrease, resulting in a decrease in velocity, counteracting the increase in velocity resulting from the decrease in bottom friction. Similarly for Inlet 2, the pressure gradient needs to increase, resulting in an increase in velocity, counteracting the decrease in velocity resulting from the increase in bottom friction. Decreasing the pressure gradient for Inlet 1 and at the same time increasing the pressure gradient over Inlet 2 is not possible when the system is identically forced at the ocean side and at the same time the water level in the basin is forced to fluctuate uniformly. As a consequence, the Escoffier Stability Model as described above does not lead to stable double inlet systems.

> It follows that a necessary, but not necessarily sufficient, condition for a set of stable cross-sectional areas is that the formulation of the hydrodynamics allows for different pressure gradients across the inlets. In practice this implies allowing for spatial variation in basin water level. This excludes the assumption of a uniformly fluctuating basin water level.

9.4 Basin with Topographic High

9.4.1 Schematization

One way to introduce spatial variations in the basin water level is by dividing the basin into two sub-basins separated by a barrier (van de Kreeke et al., 2008; de Swart and Volp, 2012). To allow for exchange, the barrier has an opening in the form of a tidal inlet. Each sub-basin is connected to the ocean by an inlet. The ocean tides off the inlets are the same. The barrier may be viewed as a topographic high which in nature could be a tidal flat. It is there where the tides entering the inlets meet, resulting in relatively low velocities and persistent sedimentation. The tidal flats, depending on extent and height, allow a certain degree of exchange between the sub-basins. For little exchange, corresponding to a small opening in the barrier, the sub-basins act independently and the double inlet system reverts to two single inlet systems. For a large opening in the barrier, the sub-basins act as a single basin with the water levels the same in both sub-basins and fluctuating uniformly.

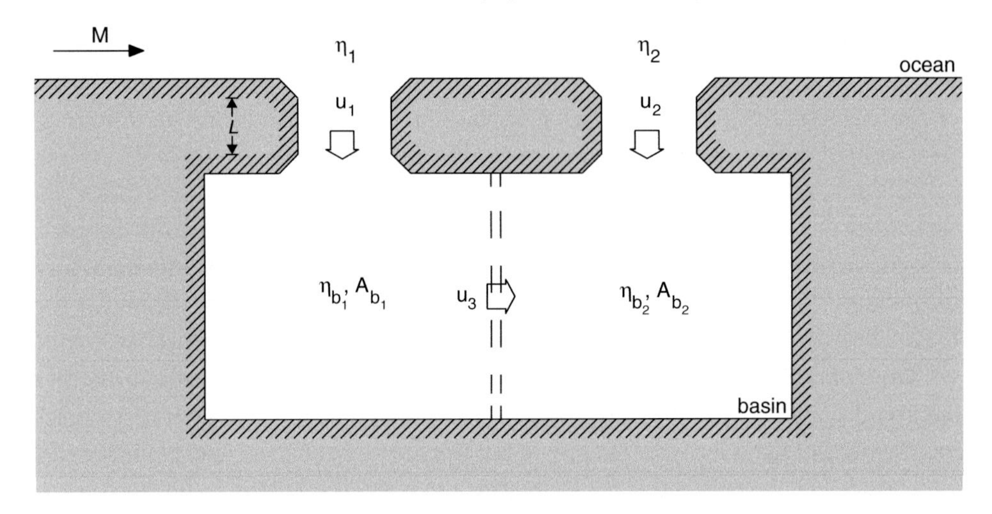

Figure 9.7 Schematization of a double inlet system with topographic high.

The tidal inlet schematization when including a topographic high is presented in Fig. 9.7. The schematization is based on the Texel-Vie inlet system. Parameter values for this system are given in Table 9.1. The topographic high divides the basin in two sub-basins with equal surface areas of 7×10^8 m^2. Assumed is a rectangular opening over the topographic high. Using Google Earth images of the Wadden Sea, the opening over the topographic high is given a width of b = 25,000 m.

9.4.2 Governing Equations

The governing equations are those used in van de Kreeke et al. (2008). The momentum equation for the flow in the inlets is Eq. (9.3). Because the inlets are relatively long, entrance/exit losses are small compared to bottom friction losses and are neglected. This result in the momentum equation

$$\frac{L_j}{g}\frac{du_j}{dt} + \frac{F_j L_j}{g\beta_2\sqrt{A_j}} u_j|u_j| = \eta_0 - \eta_{b_j}, \qquad j = 1, 2, \tag{9.8}$$

with subscript j referring to the inlets and the sub-basins. The momentum equation for the opening over the topographic high is similar to that for the inlets and reads

$$\frac{L_3}{g}\frac{du_3}{dt} + \frac{F_3 L_3 b}{g A_3} u_3|u_3| = \eta_{b_1} - \eta_{b_2}. \tag{9.9}$$

Subscript 3 refers to the opening over the topographic high. With the opening relatively wide, the hydraulic radius to a good approximation is A_3/b.

Water levels in the sub-basins are assumed to fluctuate uniformly, resulting in the continuity conditions

$$A_{b_1}\frac{\eta_{b_1}}{dt} = A_1u_1 - A_3u_3, \tag{9.10}$$

and

$$A_{b_2}\frac{\eta_{b_2}}{dt} = A_2u_2 + A_3u_3. \tag{9.11}$$

Water levels prescribed at the ocean entrance of the inlets are the same, and are given by Eq. (9.5). The equations can be solved with a finite difference method (Brouwer, 2006) or the semi-analytical method described in Appendix 2A of Brouwer (2013) and used in the present application. The semi-analytical solution leads to inlet velocities that are sinusoidal.

In Herman (2007), an exploratory model based on a similar set of equations as Eqs. (9.8)–(9.11) is used to describe the hydrodynamics of a tidal inlet system consisting of three inlets and three basins. The system of equations is solved using a finite difference method. For the same tidal inlet, a process-based simulation model is deployed, in which the hydrodynamics is described by the shallow water wave equations, eliminating the assumption of uniformly fluctuating basin water levels. The shallow water wave equations are solved numerically using the modeling system Delft3D (Lesser et al., 2004). Comparison of the results obtained from the exploratory and simulation model suggests that the exploratory model based on Eqs. (9.8)–(9.11) is capable of reproducing the water transport between basins and between basins and ocean.

9.4.3 Flow Diagrams

Using Eqs. (9.8)–(9.11) and an equilibrium velocity of 1 m s^{-1}, flow diagrams for the double inlet system are calculated for openings in the barrier with cross-sectional areas $A_3 = 25,000$ m^2, $A_3 = 50,000$ m^2 and $A_3 = 75,000$ m^2. For the relatively large opening of 75,000 m^2, the flow diagram in Fig. 9.8a shows two unstable and no stable equilibriums. The sub-basins act as one with water levels fluctuating uniformly. As a result, the pressure gradients across both inlets are always the same, thereby excluding sets of stable equilibrium cross-sectional areas. The flow diagram resembles that presented in Fig. 9.5 for a basin without a topographic high.

For the relatively small opening of 25,000 m^2, the flow diagram in Fig. 9.8c shows three unstable and one stable equilibrium. The cross-sectional areas of the stable equilibrium are $A_1 = A_2 = 83,000$ m^2. The stability of this equilibrium is explained by introducing a perturbation, whereby A_1 is increased and A_2 is decreased by 10,000 m^2, resulting in $A_1 = 93,000$ m^2 and $A_2 = 73,000$ m^2.

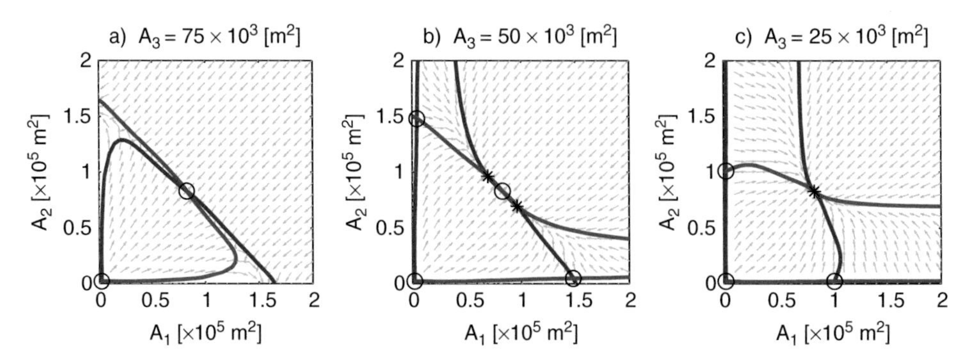

Figure 9.8 Flow diagrams for the schematized Texel-Vlie inlet system for an opening over the topographic high of a) 75,000 m^2, b) 50,000 m^2 and c) 25,000 m^2. Blue is the equilibrium velocity curve of the Texel Inlet and red is the equilibrium velocity curve of the Vlie Inlet. Circles are unstable sets and stars are stable sets of cross-sectional areas. For parameter values reference is made to Table 9.1.

Calculating the water level differences across the two inlets at times of maximum flow velocity for $(A_1, A_2) = (83,000\ \text{m}^2; 83,000\ \text{m}^2)$, results in $|\eta_0 - \eta_{b_1}| = |\eta_0 - \eta_{b_2}| = 0.071$ m. Carrying out the same calculations for the perturbed system with $(A_1, A_2) = (93,000\ \text{m}^2; 73,000\ \text{m}^2)$ results in $|\eta_0 - \eta_{b_1}| = 0.069$ m and $|\eta_0 - \eta_{b_2}| = 0.073$ m. The perturbation causes the water level difference across Inlet 1 to decrease and across Inlet 2 to increase. The decrease in the water level difference across Inlet 1 leads to a decrease in the maximum velocity from $\hat{u}_1 = \hat{u}_{eq} = 1.03$ m s^{-1} to $\hat{u}_1 = 0.99$ m s^{-1}. The increase in water level difference across Inlet 2 leads to an increase in the maximum velocity from $\hat{u}_2 = \hat{u}_{eq} = 1.03$ m s^{-1} to $\hat{u}_2 = 1.07$ m s^{-1}. As a result, Inlet 1 erodes and Inlet 2 accretes and both inlets return to the original equilibrium. The foregoing is in agreement with the statement at the end of Section 9.3 that, a necessary condition to have stable inlets is that the formulation of the hydrodynamics allows for different pressure gradients across the two inlets.

For very small openings the double inlet system approaches two single inlet systems. Assuming a complete separation of the two sub-basins and using the data in Table 9.1, the stability analysis for a single inlet system described in Chapter 8 results in stable cross-sectional areas, $A_1 = A_2 = 72,000\ \text{m}^2$. This is slightly smaller than the stable cross-sectional areas of 83,000 m^2 with the opening of 25,000 m^2, i.e., for this opening there is still some interaction between the sub-basins.

The flow diagram for the mid-size opening of 50,000 m^2 is presented in Fig. 9.8b. There are four unstable and two stable equilibriums. Although not investigated in detail, as done for the small opening, it seems reasonable to assume that also for the midsize opening pressure gradients across the two inlets can be different for the two inlets, resulting in stable equilibriums. Why there is more than one set of stable equilibriums is not clear and needs further investigation.

Contrary to the equations used by de Swart and Volp (2012), Eqs. (9.8)–(9.11) do not account for hypsometric effects. Hypsometric effects refer to the dependence of the inlet cross-sectional areas and back-barrier lagoon surface areas on the tidal stage. As shown in de Swart and Volp (2012), the conclusions when including hypsometric effects remain qualitatively the same, i.e., depending on the size of the opening over the topographic high there can be stable equilibriums. The hypsometry does affect the size of the equilibrium cross-sectional areas.

10

Cross-Sectional Stability of a Double Inlet System, Assuming a Spatially Varying Basin Water Level

10.1 Introduction

For inlets where the dynamic balance is between bottom friction and pressure gradient, stability requires spatial variations in basin water level and/or different ocean tides (Section 9.3). In a primitive way, for the basin water level this was demonstrated by introducing a topographic high. The topographic high divides the basin into two sub-basins with different uniformly fluctuating water levels. In the present chapter, instead of a topographic high, spatial variations in water level are introduced describing the hydrodynamics by the shallow water wave equations. Using these equations, the cross-sectional stability of a double inlet system is investigated. As in Chapter 9, the double inlet system resembles the Texel-Vlie inlet system but with a slightly different schematization. In the various experiments, special attention is given to the effect of basin depth, basin geometry, Coriolis acceleration and radiation damping on the spatial variation in basin water level and the cross-sectional stability. Cross-sectional stability of the inlets is determined using the Escoffier Stability Model for a double inlet system described in Section 9.2.

10.2 Schematization

The double inlet system consists of four rectangular compartments of length L_j, width B_j and uniform depth h_j ($j = 0$–3); see Fig. 10.1. Compartment 0 represents the ocean with an open boundary at $x = -L_0$. Compartments 1 and 2 are the inlets. Inlets have equal lengths and a rectangular cross-section. Compartment 3 is the tidal basin. The double inlet system, consisting of Compartments 1, 2 and 3, is symmetrically aligned with respect to the central axis of the ocean compartment. The two inlets are at equal distance, $\Delta y/2$, from the central axis.

10.3 Governing Equations and Boundary Conditions

In each compartment, conservation of momentum and mass is expressed by the depth-averaged shallow water wave equations, including the Coriolis acceleration.

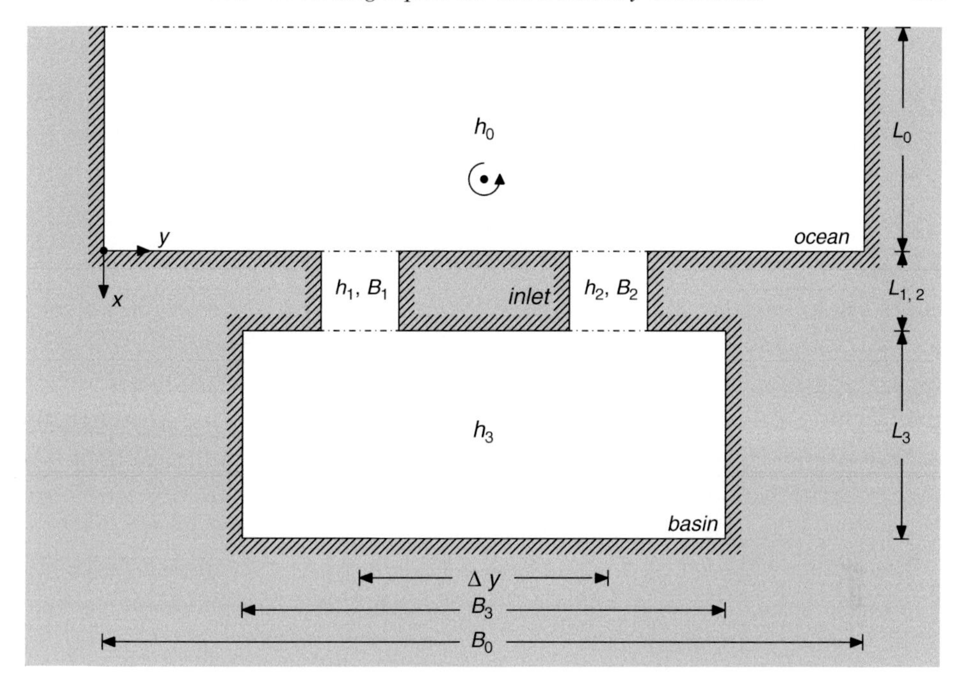

Figure 10.1 Schematized Texel-Vlie inlet system. Subscript 1 refers to Texel Inlet, subscript 2 refers to Vlie Inlet and subscript 3 refers to the basin. Furthermore, h is mean water depth and B is compartment width.

To allow for a semi-analytical solution of the equations, the advective terms are neglected and the bottom friction term is linearized. The mean water depth in all compartments is assumed to be large compared to the water level amplitude. With these constraints, the shallow water wave equations are

$$\frac{\partial u_j}{\partial t} - f v_j + \frac{r_j u_j}{h_j} = -g \frac{\partial \eta_j}{\partial x}, \tag{10.1}$$

$$\frac{\partial v_j}{\partial t} + f u_j + \frac{r_j v_j}{h_j} = -g \frac{\partial \eta_j}{\partial y}, \tag{10.2}$$

$$\frac{\partial \eta_j}{\partial t} + h_j \left[\frac{\partial u_j}{\partial x} + \frac{\partial v_j}{\partial y} \right] = 0, \tag{10.3}$$

where subscript j refers to Compartment j, with $j = 0$–3. In Eqs. (10.1)–(10.3), $u_j(x, y, t)$ and $v_j(x, y, t)$ are the depth-averaged velocities in, respectively, the x- and y-direction, $\eta_j(x, y, t)$ is the free surface elevation with respect to mean sea level, g is the gravity acceleration, h_j is the mean water depth and f is the Coriolis parameter. For the coordinate system reference is made to Fig. 10.1. The Coriolis parameter is given by

$$f = 2\Omega \sin \varphi, \tag{10.4}$$

with $\Omega = 7.292 \times 10^{-5}$ rad s^{-1} is the angular frequency of the Earth's rotation and φ is latitude. The coefficient r_j is a linear bottom friction coefficient. Using the Lorentz linearization (Lorentz, 1926; Zimmerman, 1982),

$$r_j = \frac{8FU_j}{3\pi}. \tag{10.5}$$

In Eq. (10.5), F is the nonlinear bottom friction coefficient and U_j is the velocity scale of Compartment j, defined as the velocity amplitude averaged over Compartment j. To ensure that the velocity scale agrees with the solution, the linear friction coefficients are obtained using an iterative procedure. For details, see Appendix A.4 in Brouwer et al. (2013). Even though the nonlinear bottom friction coefficient F is assumed to be the same for each compartment, the linear bottom friction coefficient differs due to the different velocity scales.

The system is forced by a single damped Kelvin wave with angular frequency σ, entering through the open boundary of the ocean compartment and propagating in the positive x-direction

$$\eta(x, y, t) = \Re\left\{ Z e^{-\frac{y}{\gamma L_r}} e^{i(\sigma t - \gamma k_0 x)} \right\}, \tag{10.6}$$

with L_r the Rossby radius of deformation for the ocean compartment defined by

$$L_r = \frac{\sqrt{g h_0}}{f}, \tag{10.7}$$

γ is a complex friction factor for the ocean compartment defined by

$$\gamma = \sqrt{1 - \frac{i r_0}{\sigma h_0}} \tag{10.8}$$

and k_0 is the wave number without bottom friction for the ocean compartment

$$k_0 = \frac{\sigma}{\sqrt{g h_0}}. \tag{10.9}$$

Z is the amplitude of the wave at $(x, y) = (-L_0, 0)$.

For the special case with the Coriolis acceleration equal to zero and introducing

$$i\gamma k_0 = \mu + ik, \tag{10.10}$$

Eq. (10.6) can be written as

$$\eta(x, y, t) = Z e^{-\mu x} \cos(\omega t - kx). \tag{10.11}$$

Eq. (10.11) represents a damped shallow water wave traveling in the positive x-direction, where μ is a damping factor and k is the wave number when accounting for bottom friction. Given k_0, σ, h_0 and r_0, μ and k follow from Eq. (10.10) (Ippen, 1966; van de Kreeke, 1998).

Using the terminology introduced in Chapter 1, the system of Eqs. (10.1), (10.2) and (10.3) with the ocean forcing given by Eq. (10.6) constitutes a process-based exploratory model. In applying the model, a water level amplitude is prescribed halfway between the two inlets. For the situation without inlets, the corresponding amplitude Z of the incoming Kelvin wave is then determined by trial and error. Due to the Coriolis acceleration, the Kelvin wave travels along the coast past the two inlets, causing the water levels off the inlets to be different. The Kelvin wave, along with other waves generated within the model domain, leaves the ocean compartment without reflecting at the open boundary.

At the closed boundaries, a no-normal flow condition,

$$u = 0 \qquad \text{and} \qquad v = 0, \tag{10.12}$$

is imposed. Continuity of elevation and normal flux is required across the interfaces between the ocean and inlet compartments, formulated as

$$\eta_0 = \eta_1 \qquad \text{and} \qquad \eta_0 = \eta_2, \tag{10.13}$$

and

$$h_0 u_0 = h_1 u_1 \quad \text{and} \quad h_0 u_0 = h_2 u_2. \tag{10.14}$$

Similarly, for the interfaces between basin and inlets holds

$$\eta_3 = \eta_1 \qquad \text{and} \qquad \eta_3 = \eta_2, \tag{10.15}$$

and

$$h_3 u_3 = h_1 u_1 \quad \text{and} \quad h_3 u_3 = h_2 u_2. \tag{10.16}$$

10.4 Solution Method

Following Brouwer et al. (2013), in each compartment the solution for η_j, u_j and v_j is written as the truncated sum of analytical wave solutions in an infinite channel. This involves incoming and reflected Kelvin waves as well as Poincaré waves generated at closed boundaries. The unknown amplitudes in the solution are determined from the matching conditions at the interfaces, Eqs. (10.13)–(10.16), and the no-normal flow condition, Eq. (10.12). For this a collocation method is used (Boyd, 2001; Roos and Schuttelaars, 2011), i.e., a large number of discrete points are defined along the interfaces and closed boundaries and the amplitude of the Kelvin and Poincaré waves are calculated such that the matching and boundary conditions in these points are satisfied. The solution leads to sinusoidal inlet velocities.

The process-based exploratory model is computationally efficient, which makes it possible to carry out large numbers of simulations in a short period of time. This makes it eminently suitable to calculate inlet velocities for a large number of combinations of cross-sectional areas (A_1, A_2), needed to construct the flow diagrams discussed in Section 9.2.6.

10.5 Effect of Spatial Variations in Basin Water Level on Cross-Sectional Stability

10.5.1 Spatial Variations in Basin Water Level

In the following a qualitative description of the effects of basin depth, Coriolis acceleration, radiation damping and basin geometry on the spatial variations of water levels is presented (Brouwer et al., 2013).

Basin depth. Finite basin depth affects the tidal propagation and dissipation. The tidal wave is short compared to the same wave in deep water, resulting in differences in surface elevation over relatively short distances.

Coriolis acceleration. The Coriolis acceleration leads to asymmetry and thus spatial variation in the basin surface elevation. In addition, the Coriolis acceleration allows prescribing the ocean forcing by a Kelvin wave traveling along the coastal boundary, resulting in different ocean tides off the inlets.

Radiation damping. In earlier studies, the forcing of the system has been prescribed by water levels, usually in the form of a series of tidal components. Formulating the boundary conditions in this way does not account for the effect of the inlets on the forcing. This effect is included by prescribing the boundary forcing by an incoming wave. The resulting water level variations off the inlets induce oscillating flow in each inlet which triggers co-oscillations in the basin. In turn, these co-oscillations result in waves radiating away into the ocean, influencing the surface elevation off the inlets. This mechanism is referred to as radiation damping. The effect of radiation damping on the surface elevations off the inlets decreases with decreasing values of the ratio of ocean and inlet depth (Buchwald, 1971).

Basin geometry. Short and wide basins, as opposed to long and narrow basins, make a difference in the spatial variation of basin water level.

The effect of basin water depth, Coriolis acceleration, radiation damping and basin geometry on the basin water level and inlet stability is further investigated using the hydrodynamics model described in Sections 10.3 and 10.4 and the Escoffier Stability Model introduced in Section 9.2. In the experiments the water level amplitude described halfway between the inlets is $Z_{\mathrm{char}} = 0.8$ m.

10.5.2 Comparison with Earlier Stability Analysis

As a first step, the process-based exploratory model is used to evaluate the stability of the schematized Texel-Vlie inlet system described in Section 9.2. Parameter values are presented in Table 10.1 and are close to the same as those presented in Table 9.1. To compare with the stability analysis of the Texel-Vlie inlet system in Section 9.2, the basin depth is given a large value $h_3 = 1{,}000$ m, resulting in a close to uniformly fluctuating basin water level. In the stability analysis the inlet cross-sections are assumed to be geometrically similar, resulting in $h_j = 0.07\sqrt{A_j}$ with $j = 1, 2$ (Appendix 8.A), with subscript 1 referring to the Texel Inlet and subscript 2 referring to the Vlie Inlet. To minimize radiation damping, the ocean compartment is given a large depth, $h_0 = 1{,}000$ m. The basin dimensions are 30 × 40 km. With the Coriolis parameter equal to zero, the Rossby radius of deformation

Table 10.1 *Parameter values for schematized Texel-Vlie inlet system.*

Parameter	Symbol	Dimension	Value
Ocean compartment (0)[a]			
Length	L_0	km	50
Width	B_0	km	100
Inlet compartment (1)[b]			
Length	L_1	km	6
Inlet compartment (2)[b]			
Length	L_2	km	6
Basin compartment (3)[a]			
Length	L_3	km	30
Width	B_3	km	40
General			
Distance between inlets	Δy	km	10
Ocean water level amplitude	$\hat{\eta}_0$	m	0.8
Tidal wave frequency (M_2)	σ	rad s^{-1}	1.4×10^{-4}
Bottom friction factor	F	–	2.5×10^{-3}
Equilibrium velocity	$\hat{u}_{eq}$	m s^{-1}	1.0
Shape factor[b]	β_2	–	0.07

[a]Compartment depth varies depending on experiment.
[b]Inlet cross-sections are rectangular with a width to depth ratio of 200, resulting in a shape factor with a value of 0.07 (Appendix 8.A).

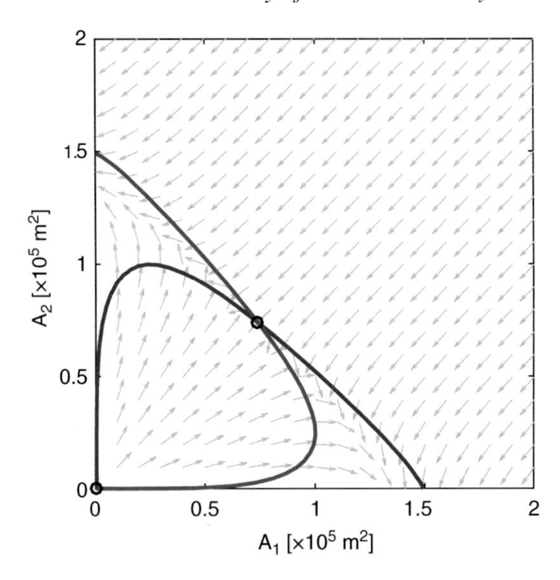

Figure 10.2 Flow diagram for the schematized Texel-Vlie inlet system; basin depth $h_3 = 1,000$ m, ocean depth $h_0 = 1,000$ m, basin dimensions 30×40 km, Coriolis parameter $f = 0$. Circles represent sets of unstable cross-sectional areas. Blue is equilibrium velocity curve for Inlet 1 and red is equilibrium velocity curve for Inlet 2. (From: Brouwer et al., 2013)

L_r approaches infinity and the prescribed Kelvin wave reduces to a damped shallow water wave traveling perpendicular to the coast. This results in the same tides off both inlets. The resulting flow diagram is presented in Fig. 10.2. The same as in Fig. 9.6, the flow diagram shows two unstable and no stable equilibriums. Minor differences are attributed to differences in the shape factor and the equilibrium velocity; compare parameter values in Tables 9.1 and 10.1.

10.5.3 Effects of Basin Depth, Coriolis Acceleration, Radiation Damping and Basin Geometry

To demonstrate the effects of basin depth, Coriolis acceleration, radiation damping and basin geometry on the stability of a double inlet system, four experiments are carried out (see Fig. 10.3). The schematized Texel-Vlie inlet system with the parameter values presented in Table 10.1 is used as the inlet configuration.

Basin Depth

To investigate the effect of basin depth on cross-sectional stability, the basin is given a shallow depth $h_3 = 5$ m. With the Coriolis parameter $f = 0$, the incoming wave is a damped shallow water wave propagating perpendicular to

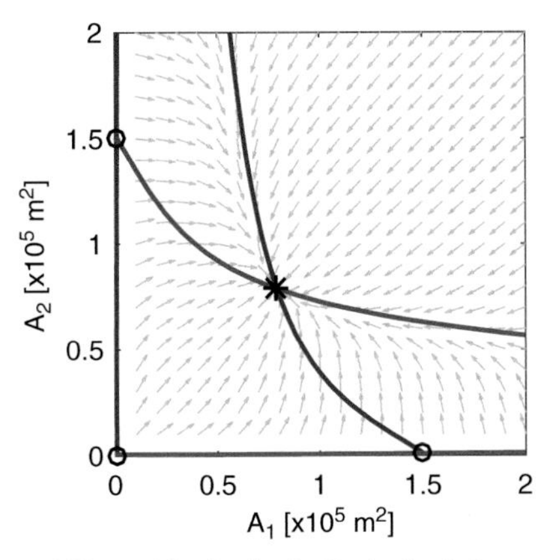

(a) **Effect of basin depth:** Basin depth $h_3 = 5$ m, ocean depth $h_0 = 1,000$ m, basin dimensions 30 × 40 km, Coriolis parameter $f = 0$.

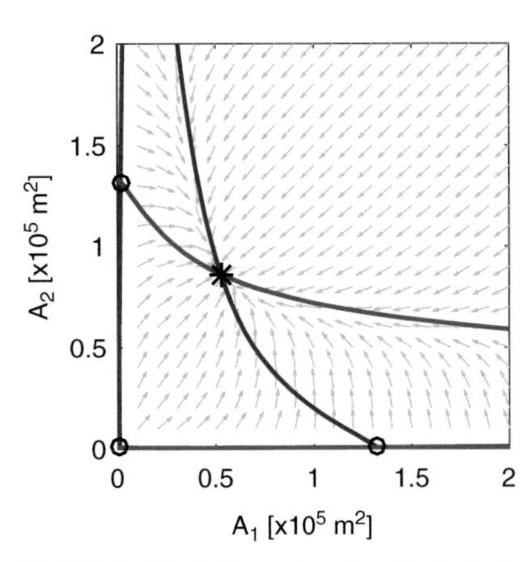

(b) **Effect of Coriolis acceleration:** Basin depth $h_3 = 5$ m, ocean depth $h_0 = 1000$ m, basin dimensions 30 × 40 km, Coriolis parameter $f = 1.164 \times 10^{-4}$ rad s^{-1}.

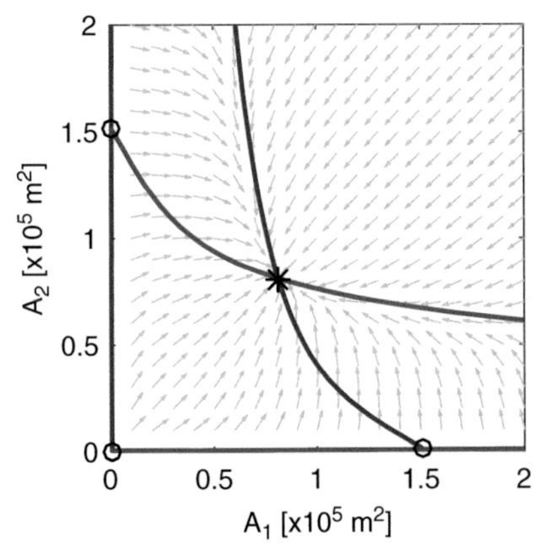

(c) **Effect of radiation damping:** Basin depth $h_3 = 5$ m, ocean depth $h_0 = 20$ m, basin dimensions 30 × 40 km, Coriolis parameter $f = 0$.

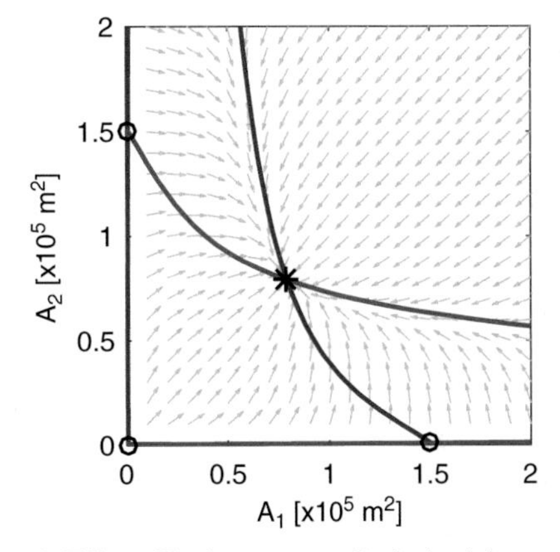

(d) **Effect of basin geometry:** Basin depth $h_3 = 5$ m, ocean depth $h_0 = 1000$ m, basin dimensions 60 × 20 km, Coriolis parameter $f = 0$.

Figure 10.3 Flow diagrams for the schematized Texel-Vlie inlet system displaying the effect of a) basin depth, b) Coriolis acceleration, c) radiation damping and d) basin geometry. Circles are unstable sets and stars are stable sets of cross-sectional areas. Blue is equilibrium velocity curve for Inlet 1 and red is equilibrium velocity curve for Inlet 2.

the coast. This results in equal tides at the inlets. The water depth in the ocean compartment is $h_0 = 1,000$ m. This is relatively large compared to the depth of the inlets, thus limiting the effect of radiation damping on the offshore water levels. The basin dimensions are 30 × 40 km. The flow diagram, presented in Fig. 10.3a, shows four equilibriums, one of which is stable and the other three are unstable.

Coriolis Acceleration

To investigate the effect of the Coriolis acceleration on the cross-sectional stability, the Coriolis parameter is given a value $f = 1.164 \times 10^{-4}$ rad s^{-1}, corresponding to the latitude of the Texel-Vlie inlet system. Other parameters remain the same as in the experiment with the finite basin depth, i.e., $h_0 = 1,000$ m and $h_3 = 5$ m. The basin dimensions are 30 × 40 km. As a result of the Coriolis acceleration, the incoming wave is a Kelvin wave traveling along the coast, resulting in different amplitudes and phases of the ocean tides at the inlets. In addition to the water levels off the inlets, the Coriolis acceleration affects the basin water levels, leading to asymmetry in the spatial pattern of the basin tidal amplitudes. The flow diagram when including the Coriolis acceleration is presented in Fig. 10.3b. The flow diagram differs little from that presented in Fig. 10.3a. This suggests that the effect of the Coriolis acceleration on the cross-sectional stability is of secondary importance compared to that of basin depth. As a result of the Coriolis acceleration, the problem becomes asymmetric and the stable cross-sectional areas differ slightly for the two inlets.

Radiation Damping

The effect of radiation damping on cross-sectional stability is investigated by decreasing the depth of the ocean compartment to $h_0 = 20$ m, thereby increasing the effect of radiation damping. The basin depth remains $h_3 = 5$ m. The basin dimensions are 30 × 40 km. The Coriolis parameter $f = 0$ and the incoming wave is a damped shallow water wave traveling perpendicular to the coast, resulting in the same tides off both inlets. The flow diagram is presented in Fig. 10.3c and shows three unstable and one stable equilibrium. The equilibrium cross-sectional areas differ little from those in Fig. 10.3a, showing the marginal effect of radiation damping on the cross-sectional stability compared to that of reducing the basin depth.

Basin Geometry

To demonstrate the effect of basin geometry on cross-sectional stability, the basin is given a length $L_3 = 60$ km and a width $B_3 = 20$ km (as opposed to 30 × 40 km in the previous experiments). The ocean depth $h_0 = 1,000$ m, the Coriolis parameter $f = 0$ and the basin depth $h_3 = 5$ m. The incoming wave travels perpendicular to the coast. This results in the same tides off both inlets. The flow diagram, presented in Fig. 10.3d, is qualitatively the same as that in Fig. 10.3a with four equilibriums, one of which is stable and the other three are unstable. However, the stable cross-sectional areas are now considerably smaller.

In summary, allowing for a spatially varying basin water level by introducing a finite basin depth results in three sets of unstable and one set of stable equilibrium cross-sectional areas. The same result is found when, in addition to finite basin depth, Coriolis acceleration, radiation damping and different basin geometry is included. Accounting for Coriolis acceleration and radiation damping slightly affects the size of the stable equilibrium cross-sectional areas. Changes in basin geometry significantly affect the size of the stable equilibrium cross-sectional areas.

10.6 Multiple Inlets

In a recent study, the stability analysis for the double inlet system was expanded to include multiple inlets (Roos et al., 2013). In search for multiple stable inlets, their simulations start with a large number of open inlets along a barrier island coast. During the simulations some inlets close, while others remain open competing for the remaining tidal prism. The final result is a barrier island coast with a number of stable inlets. It was found that for a given length of barrier island coast there is a maximum number of stable inlets. Furthermore, simulations revealed that the number of stable inlets increases with increasing values of tidal range and basin surface area and decreasing values of longshore sand transport.

11

Morphodynamic Modeling of Tidal Inlets Using a Process-Based Simulation Model

11.1 Introduction

Morphodynamic models come in two categories. Based on their architecture a distinction is made between process-based models and empirical models. In the present chapter the focus is on process-based modeling. Process-based models start with small-scale physics and integrate the results over the larger timescales. Because at this time the state of the art of process-based modeling limits the time period over which can be integrated, most model applications focus on problems with timescales of months to decades. Examples are the adaptation of the inlet cross-section after a storm, the formation of an ebb delta after the opening of a new inlet, migration and breaching of channels and spit formation (Nahon et al., 2012; Tung et al., 2012; van der Wegen et al., 2010). To address problems with larger timescales, parallel to the process-based models, empirical models have been developed. They are the subject of Chapter 12. A common characteristic of process-based and empirical morphodynamic models is the feedback between morphology (bathymetry) and hydrodynamics; the hydrodynamics causes a change in bathymetry and in turn this affects the hydrodynamics.

11.2 Model Concept and Formulation

Process-based morphodynamic models consist of a series of computational modules as shown in Fig. 11.1. Starting with a known bathymetry and water level boundary conditions, the hydrodynamic equations (Lesser et al., 2004) are solved in the Flow module. Given the wave boundary conditions, wave transformation including wave height and direction is calculated in the Waves module. To account for tidal stage and wave-current interaction, information on water levels and current velocities is transferred from the Flow to the Waves module. Radiation stresses are calculated and the results are transferred to the Flow module. Exchange between the two modules takes place at specified time intervals.

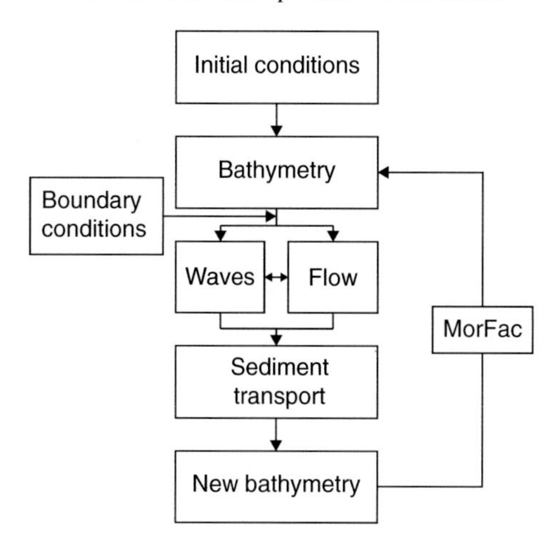

Figure 11.1 Example of an elementary feedback loop.

In general this interval is much larger than the computational time step used in either of the modules. Using the information on currents and waves from the Flow and Waves modules, sediment transport is calculated in the Sediment Transport module using selected sediment transport formulae. With the results of the sand transport calculations the bathymetry is updated in the Bathymetry module.

The computer time for one complete cycle varies with the application and is measured in seconds to minutes. Even though short, for calculations extending over a large time period, e.g., one year, this would still lead to lengthy computer times. Computer time is reduced by taking advantage of the difference in timescales of the hydrodynamic and morphodynamic processes. For this, the depth changes after one cycle are multiplied by a morphological factor MorFac (Roelvink and Reniers, 2010). As an example, assuming MorFac is 30 and the computer time of one complete cycle is one minute, the change in bathymetry over a period of 30 minutes is obtained, while wave, current and transport fields only have to be calculated once during that one-minute period. The wave height and direction, and current and transport fields, only have to be calculated once during that 30-minute period. Selecting the value of MorFac is not trivial, as too large a value can lead to the wrong morphology.

As examples of process-based morphodynamic modeling, two sets of experiments presented in Tung et al. (2012) are discussed. The purpose of the first set of experiments is to simulate the development of the inlet morphology after the opening of a new inlet. The purpose of the second set of experiments is to investigate the cross-sectional area – tidal prism relationship for a set of geologically and hydrodynamically similar inlets.

11.3 Morphology of a Newly Opened Inlet

In the calculations, the depth-averaged version of the Delft3D online modeling system is used (Lesser et al., 2004). Following Murray (2003), this system can be characterized as a process-based simulation model. In the present application, in the Flow module the water motion is described by the shallow water wave equations, excluding the Coriolis term. The Waves module uses the Swan wave model (Booij et al., 1999). At intervals of 30 minutes, information on water levels and current velocities is transferred from the Flow to the Waves module. In the Sediment Transport module the sediment transport formulation of van Rijn (1993) is used, which includes both tide- and wave-induced sediment transport. The formulation accounts for bottom sediment transport and suspended load transport. The model includes dry bank and flat erosion. Carrying out experiments in which the morphological factor MorFac was varied systematically, it was found that for values of 50 and less the model results are unaffected by the upscaling. A value of 40 was selected for the numerical experiments in this study. The horizontal eddy viscosity coefficient and eddy diffusion coefficient are 0.1 m^2 s^{-1}. The Chézy friction coefficient is the same for the entire computational domain and equal to 65 $m^{1/2}$ s^{-1}. Bottom sediment consists of a single fraction of non-cohesive sand with a density of 2,650 kg m^{-3} and a median grain diameter of 250 μm.

The idealized tidal inlet system consists of a rectangular basin connected to the ocean by an inlet. Initially, the tidal basin and inlet have a uniform depth; the cross-section of the inlet is trapezoidal. Inside the surf zone the bathymetry has a concave equilibrium profile (Dean, 1991). Outside the surf zone a gentle profile with a slope 1:200 to a water depth of SWL -13 m (still water level) is used. The elevation of the barrier islands on both sides of the inlet channel is set at SWL +3 m. To allow for an unrestricted widening and/or migration of the inlet channel the barrier islands are defined as erodible barriers.

The computational grid is rectangular with a resolution of 30 m in the inlet region, gradually increasing to 200 m in the offshore and basin area. A simple harmonic tide is prescribed at the ocean boundary, located 3 km offshore. The lateral boundaries are open, non-reflective Neumann boundaries where a zero alongshore water level gradient is prescribed (Roelvink and Walstra, 2004). Waves in the form of a JONSWAP spectrum are prescribed at the offshore boundary. The spectrum is characterized by a significant wave height, peak period and mean wave direction.

To study the development of the morphology of a newly opened inlet, experiments were carried out with two different forcings: one with tide only and the other with tides and waves. Parameter values used in the experiments are listed in Table 11.1. With waves approaching the shore at an oblique angle, the major difference between the two experiments is the absence of longshore sand transport for tide only. For the experiment with waves, the calculated longshore sand

Table 11.1 *Morphology of a newly opened inlet; parameter values used in the experiments.*

	Parameter	Value
Basin	Surface area A_b	15 km^2
	Initial depth H_0	2 m
Inlet	Length L	500 m
	Initial depth H_0	2 m
	Shape cross-section	trapezoidal
Tide	Amplitude $\hat{\eta}_0$	0.5 m
	Period T	12 hours
Waves	Significant wave height H_s	1.5 m
	Peak frequency f_p	7 s
	Direction from north ϑ	25°
Sand	Median diameter d_{50}	250 μm
	Density ρ_s	2650 kg m^{-3}
MorFac	MF	40

transports is 0.5×10^6 m^3 year^{-1}. Except for the waves and the initial cross-sectional area of the inlet (850 m^2 for the experiment with tide only and 250 m^2 for the experiment with waves and tide), parameter values are the same for both experiments. The main interest here is to determine whether the morphology approaches equilibrium and, if so, what the equilibrium conditions are. To highlight the difference in response, results of the two experiments are presented in parallel.

The initial bathymetry and the bathymetry after 10,000 days for tide only and 800 days for waves and tide are presented in Fig. 11.2. The reason for the relatively short simulation time for the experiment with tide and waves is the much faster morphological response in comparison to the experiment with tide only. Comparing Figs. 11.2a and 11.2b, for tide only the bathymetry remains symmetric with the inlet perpendicular to the shore. Comparing Figs. 11.2c and 11.2d, when adding waves, and thus longshore sand transport, the bathymetry becomes asymmetric with the inlet taking on a NE–SW direction. The P/M ratio in the experiment is approximately 50, corresponding to poor to fair location stability (Section 3.5). In agreement with this low value, the inlet has shifted in the downdrift direction.

Changes in the cross-section at the gorge for tide only and tide and waves are presented in Figs. 11.3a and 11.3b, respectively. For tide only, the cross-section maintains its position. The width at mean sea level remains the same, but the depth increases resulting in a change from the original trapezoidal to a V-shaped cross-section. For tide and waves, the cross-section shifts in the downdrift direction. The width at mean sea level increases and so does the depth. Furthermore, there is a tendency to form a dual channel system.

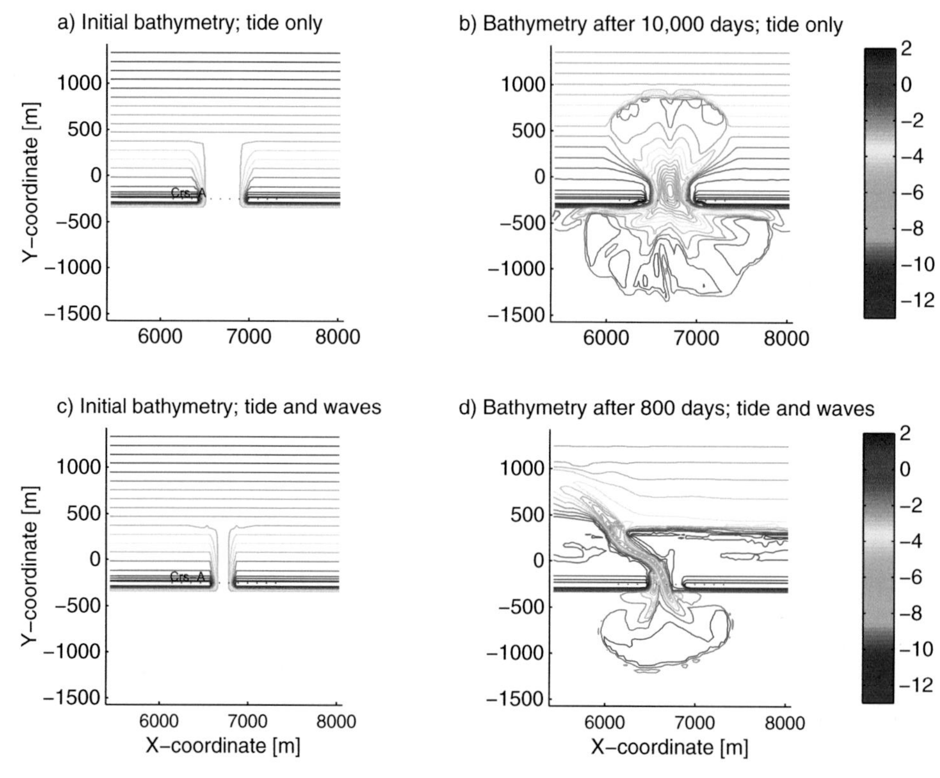

Figure 11.2 a) Initial bathymetry for the experiment with tide only, b) bathymetry for tide only after 10,000 days, c) initial bathymetry for the experiment with tide and waves, and d) bathymetry for tide and waves after 800 days (figures a and b reprinted from Tung et al., 2012, copyright 2016, with permission from Elsevier). For parameter values reference is made to Table 11.1.

Tidal prism P, velocity amplitude $\hat{u}$ and cross-sectional area A as a function of time are presented in Fig. 11.4a for tide only and in Fig. 11.4b for tide and waves. The tidal prism was calculated by integrating the cross-sectionally averaged velocity over the ebb cycle. With the known tidal prism, the velocity amplitude follows from Eq. 5.6. For tide only, the velocity amplitude approaches a value of 0.5 m s^{-1}. This is considerably larger than the critical velocity of erosion of 0.27 m s^{-1} for the sand used in the experiments; the sand bottom constitutes a 'live bed' where sand carried in by the flood is removed during ebb. For tide and waves, the equilibrium velocity is even larger and approaches a value of 0.8 m s^{-1}. This is attributed to the wave-induced longshore sand transport (Section 5.2.4). Similar to the velocity amplitude, the cross-sectional areas for both tide only and tide and waves approach an equilibrium value. The larger equilibrium cross-sectional area corresponds with the smaller equilibrium velocities. This agrees with the Escoffier Diagram in Section 8.2.2.

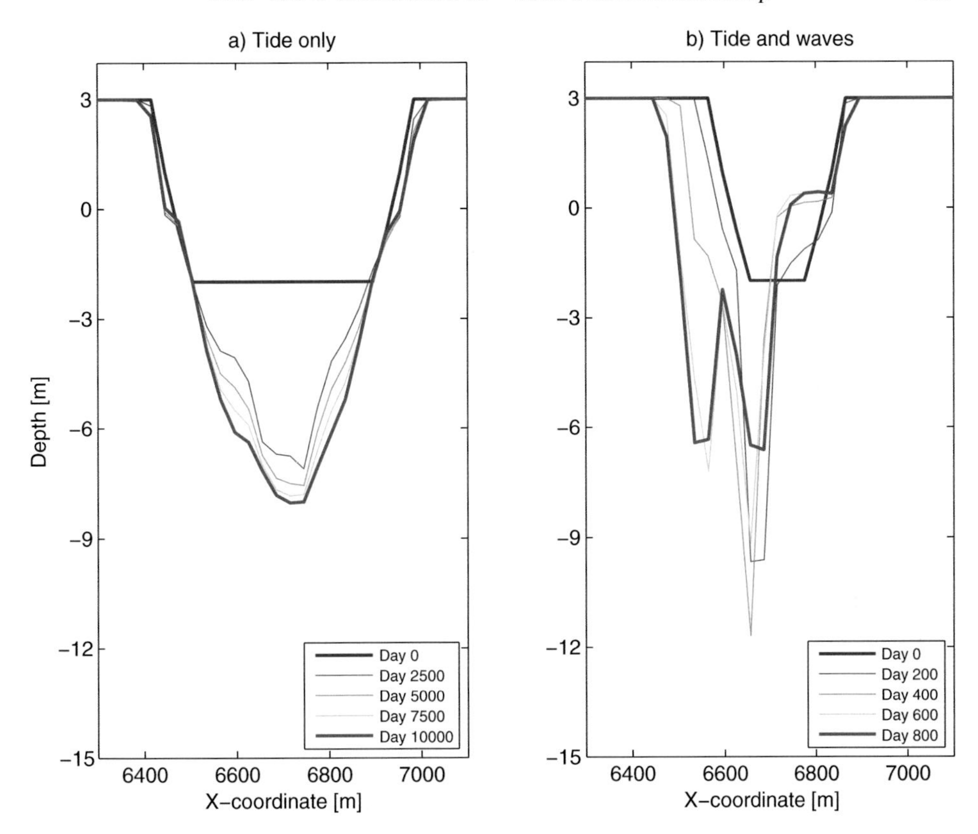

Figure 11.3 Changes in cross-section at the gorge for a) tide only and b) tide and waves. For parameter values and coordinate system, reference is made to Table 11.1 and Fig. 11.2, respectively.

To investigate the dependence of the equilibrium cross-sectional areas and equilibrium velocity on the value of the initial cross-sectional area, numerical experiments were carried out with initial cross-sectional areas that are smaller and larger than the estimated equilibrium values. The velocity amplitudes for initial cross-sectional areas of 250 m^2 and 2,400 m^2 are presented in Fig. 11.5. Regardless of the initial value, the cross-sectional areas approach the same equilibrium value and the same holds for the velocity amplitude. Additional examples are presented in Tung et al. (2012). The cluster of points near the equilibrium is an indication that the equilibrium is dynamic rather than static.

11.4 Cross-Sectional Area – Tidal Prism Relationship

Using the depth-averaged version of the Delft3D online modeling system, the cross-sectional area – tidal prism relationship for inlets at equilibrium is determined

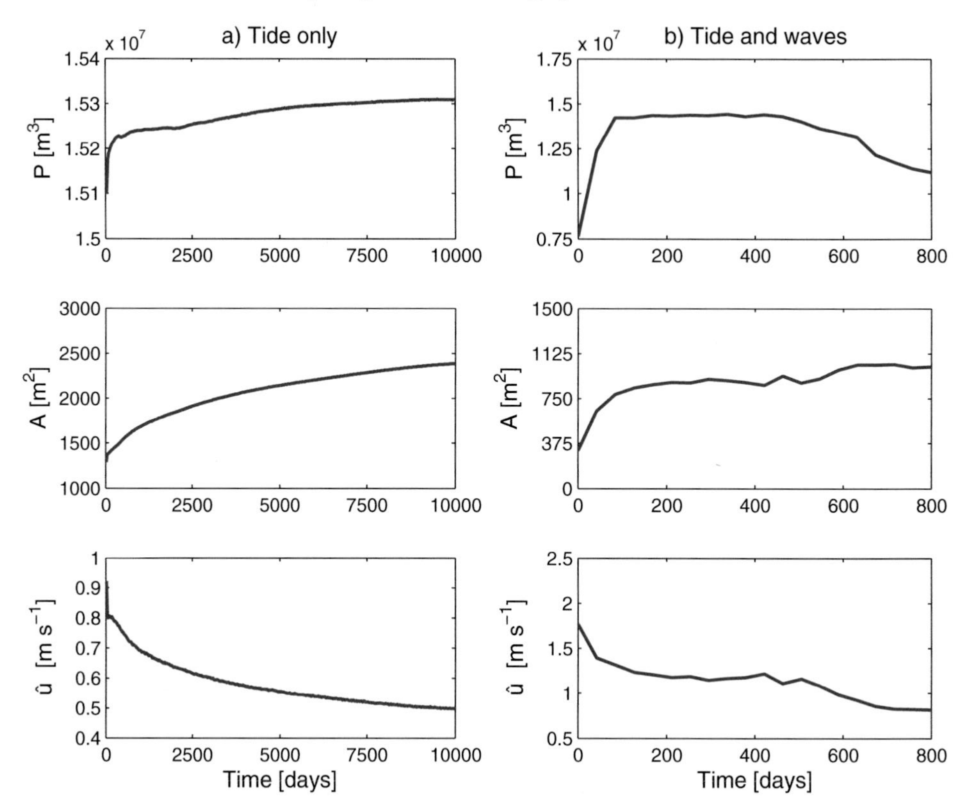

Figure 11.4 Tidal prism, inlet cross-sectional area and velocity amplitude for a) tide only and b) tide and waves. For parameter values reference is made to Table 11.1.

for a set of five geologically and hydrodynamically similar inlets. Experiments were carried out for different values of ocean tidal amplitude, a_0, and basin surface area, A_b, as summarized in Table 11.2. The remaining parameter values are kept constant, i.e., tidal period is 12 hr, offshore significant wave height is 0.7 m, peak wave period is 6 s, wave direction is 25° from the north, median grain diameter is 250 μm, grain density is 2,650 kg m^{-3}, initial depth of the basin and the inlet is 2 m. The equilibrium cross-sectional areas and corresponding velocity amplitudes are calculated using initial cross-sectional areas that are smaller and larger than the estimated equilibrium cross-sectional area. The selected vales are 250 m^2 and 2,400 m^2. The resulting equilibrium cross-sectional areas together with the corresponding tidal prisms are presented in the last two columns of Table 11.2. The longshore sand transport calculated with the model is the same for all experiments and equals 50,000 m^3 year^{-1}.

Table 11.2 *Results of model experiments used to determine the cross-sectional area – tidal prism relationship (Tung et al., 2012).*

Experiment	a_0 [m]	A_b [m^2]	A [m^2]	P [m^3]	$\hat{u}$ [m s^{-1}]
1	0.50	15×10^6	1,380	15.3×10^6	0.81
2	0.75	15×10^6	2,400	22.7×10^6	0.69
3	1.00	15×10^6	3,450	29.8×10^6	0.63
4	0.60	30×10^6	3,560	37.5×10^6	0.77
5	0.30	15×10^6	1,080	9.3×10^6	0.63

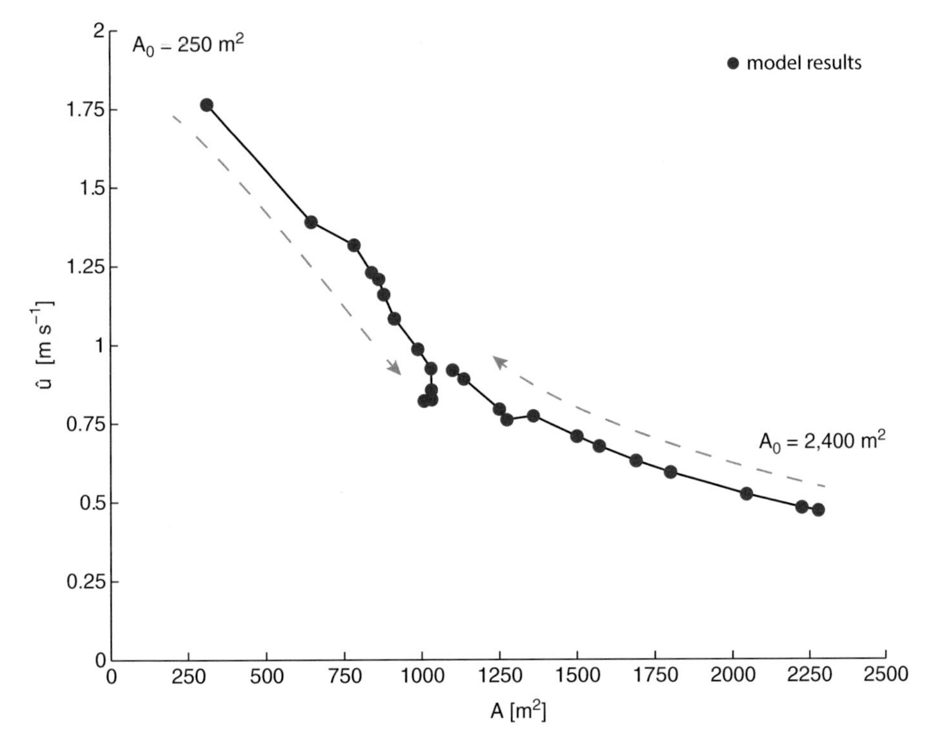

Figure 11.5 Velocity amplitude $\hat{u}$ versus cross-sectional area A for two values of the initial cross-sectional area A_0. For parameter values, reference is made to Table 11.1.

Values of equilibrium cross-sectional areas and tidal prisms are plotted in Fig. 11.6. Using the correlation functions, Eqs. (5.1) and (5.2), a best fit results in, respectively,

$$A = 2.33 \times 10^{-4} P^{0.95} \qquad \text{with} \quad r^2 = 0.96 \tag{11.1}$$

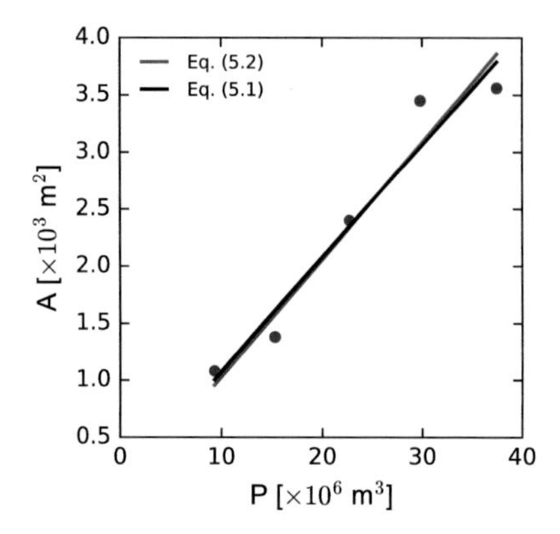

Figure 11.6 The cross-sectional area – tidal prism relationship for five experiments in Table 11.2.

and

$$A = 1.03 \times 10^{-4} P \qquad \text{with} \quad r^2 = 0.94. \tag{11.2}$$

Similar to the inlets of the Dutch Wadden Sea and the North Island of New Zealand, the equilibrium cross-sectional areas calculated with the morphodynamic model show good correlation with the tidal prisms. It is noted that the coefficient in Eq. (11.2) is considerably larger than for the inlets of the Dutch Wadden Sea and the North Island of New Zealand (Section 5.2). Referring to Eq. (5.7), and with k values approximately the same, this is attributed to the small longshore sand transport of 50,000 m^3 year^{-1} in the morphodynamic model.

11.5 Limitations of Process-Based Morphodynamic Models

The examples presented in Sections 11.3 and 11.4 illustrate the potential of modeling morphological processes in tidal inlets. Although the results look promising, process-based morphodynamic models are still in a developmental state. In particular, there are fundamental problems associated with the upscaling of the small-scale processes to larger space and timescales resulting from (1) limited knowledge of sediment transport processes, (2) the statistical nature of the wave forcing, (3) the propagation of numerical errors when integrating over large timescales (larger than decades) and (4) computer time. The formulation of the sediment transport by tidal currents and waves is still largely empirical. Wave forcing is only known in a statistical sense, whereas indications are that morphological changes are sensitive to the time history of the forcing (Southgate, 1993). With regards to the propagation of numerical errors, there is the problem that small changes in initial conditions can ultimately lead to widely different results (Ridderinkhof and Zimmerman, 1992).

As a result of the strongly varying bathymetry, and to obtain sufficient spatial resolution, the space step in the calculations has to be small. As a consequence, to maintain computational stability, the time step has to be small. For large simulation periods, this leads to lengthy computing time. Fortunately, with the development of faster computers this becomes less and less of a problem.

In view of the aforementioned shortcomings, the general belief is that, at this time, upscaling of small-scale processes to larger timescales can only be done over short periods, after which everything diverges and collapses; indeed, this behavior has been seen in many cases when a model that was calibrated over a period of some years was continued for a much longer period (Roelvink and Reniers, 2010). Recently, a number of improvements have been introduced that allow the process-based morphodynamic models to be run over longer periods and reduce computer time. A key issue in this has been the strategy to bridge the gap between short-term hydrodynamic processes, with timescales of hours to days, and morphological changes, with timescales of months and longer. A number of strategies for this are presented in Roelvink (2006). The most promising seems to be the morphological factor or online approach used in the present application. In spite of progress, it is concluded that at this time process-based morphodynamic models should only be used to diagnose, as opposed to predict, the morphological behavior of inlets. For the use of process-based morphodynamic models in a predictive mode, additional research and development is needed. As part of this research and development effort, process-based morphodynamic models have been applied to schematized tidal inlets. Examples are found in van der Wegen et al. (2010), Nahon et al. (2012), and Tung et al. (2012). In van der Wegen et al. (2010), special attention is given to morphological changes covering timescales of centuries.

12

Morphodynamic Modeling of Tidal Inlets Using an Empirical Model

12.1 Introduction

Process-based models are a valuable tool when the relevant timescales are measured in months to years (Section 11.5). For timescales of decades to centuries, recourse is often taken to empirical models, also referred to as long-term behavior models or aggregate models. Empirical models start with the premise that, after a perturbation, the morphology tends towards an equilibrium state. The equilibrium is defined by equations of state. Examples are the relationship between inlet cross-sectional area and tidal prism (Eqs. (5.1) and (5.2)) and the relationship between ebb delta volume and tidal prism (Eq. (5.14)). Examples of perturbations are changes in the inlet morphology resulting from a storm and changes resulting from such engineering activities as dredging and basin reduction. The objective of empirical modeling is to predict the transition from the old to the new equilibrium.

12.2 Modeling Concepts

In applying empirical modeling to tidal inlets, the morphology is divided into a number of large-scale geomorphic elements, e.g., inlet, ebb delta and flood delta. Each of these elements is viewed on an aggregated scale and characterized by either a sand or a water volume. When the morphology as a whole tends to an equilibrium, so do the individual elements.

Examples of empirical models are presented in Kraus (2000), van de Kreeke (2006) and Stive et al. (1998). Kraus (2000) applied an empirical model to simulate the ebb delta development after the opening of Ocean City Inlet (MD). van de Kreeke (2006) used a similar empirical model to explain the transition from the old to the new equilibrium of the Frisian Inlet (The Netherlands) after basin reduction. In both studies, the sand transport entering and leaving an element is prescribed in terms of the ratio between the actual and equilibrium sand or water volume of the element. The model by Stive et al. (1998) is fundamentally different from the models used by Kraus (2000) and van de Kreeke (2006) in that the sand transport entering and leaving an element is formulated as a diffusive transport.

12.3 Ebb Delta Development at Ocean City Inlet

12.3.1 Ocean City Inlet

Ocean City Inlet (MD) was opened in 1933 by a hurricane. Since that time, the development of the ebb delta has been observed at regular intervals. Starting at the updrift side of the inlet, the ebb delta is divided in three geomorphic features: an ebb tidal shoal, a bypassing bar and an attachment bar; see Fig. 12.1. Sand is transported from the updrift coast to the ebb tidal shoal and from there continues its path via the bypassing bar and attachment bar to the downdrift coast. Little information is available on the hydrodynamic setting of the inlet.

12.3.2 Schematization and Model Formulation

The empirical model, referred to in Kraus (2000) as the reservoir model, consists of three elements, each representing one of the three ebb delta elements. In the model the elements are connected in series. Each element is characterized by a sand volume, V. The elements together with sand transport pathways are presented in Fig. 12.2.

Sand transport entering an element equals the sand transport leaving the neighboring updrift element. Referring to Fig. 12.2, the sand transport entering Element 1 (ebb tidal shoal) is taken as being equal to the longshore sand transport M,

$$S_{e_1} = M, \tag{12.1}$$

where the subscript e stands for entering and the subscript 1 refers to Element 1. The sand transport leaving Element 1 is taken as being proportional to the product of the sand transport entering and the ratio of the actual and equilibrium sand volume

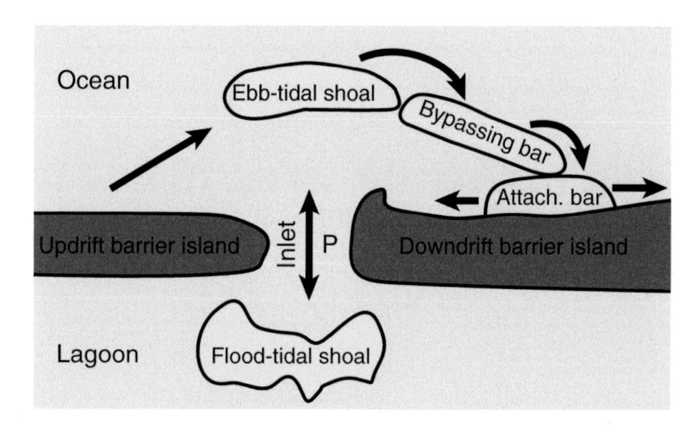

Figure 12.1 Geomorphic elements and sand transport pathways at Ocean City Inlet (MD) (adapted from Kraus, 2000, with permission from ASCE).

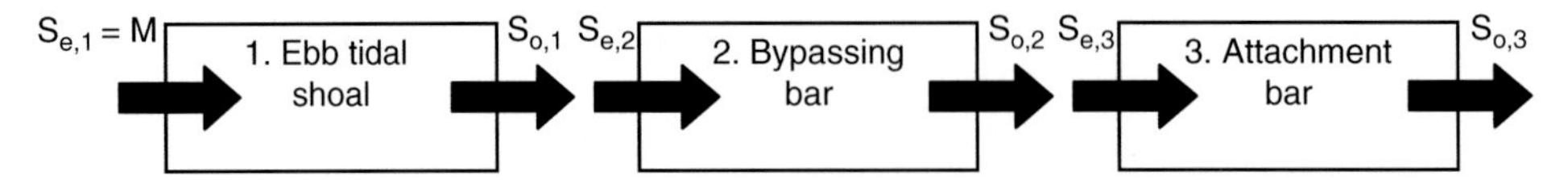

Figure 12.2 Empirical model for Ocean City Inlet; for definition of symbols reference is made to the text.

$$S_{l_1} = M\frac{V_1}{V_{1_{eq}}}, \tag{12.2}$$

where the subscript l stands for leaving, V_1 is the actual sand volume and $V_{1_{eq}}$ is the equilibrium sand volume.

The sand transport entering Element 2 (bypassing bar) equals the sand transport leaving Element 1,

$$S_{e_2} = S_{l_1}. \tag{12.3}$$

The sand transport leaving Element 2 is taken as being equal to the product of the sand transport entering and the ratio of the actual and equilibrium sand volume,

$$S_{l_2} = M\frac{V_1}{V_{1_{eq}}}\frac{V_2}{V_{2_{eq}}}. \tag{12.4}$$

Similarly, the sand transport entering Element 3 (attachment bar) is

$$S_{e_3} = S_{l_2}, \tag{12.5}$$

and the sand transport leaving Element 3 is

$$S_{l_3} = M\frac{V_1}{V_{1_{eq}}}\frac{V_2}{V_{2_{eq}}}\frac{V_3}{V_{3_{eq}}}. \tag{12.6}$$

With the sand transports defined by Eqs. (12.1) and (12.2), the sand conservation equation for Element 1 is

$$\frac{dV_1}{dt} = M\left(1 - \frac{V_1}{V_{1_{eq}}}\right). \tag{12.7}$$

Similarly, using Eqs. (12.3) and (12.4) and Eqs. (12.5) and (12.6), the sand conservation equation for Element 2 and Element 3 are, respectively,

$$\frac{dV_2}{dt} = M\frac{V_1}{V_{1_{eq}}}\left(1 - \frac{V_2}{V_{2_{eq}}}\right), \tag{12.8}$$

and

$$\frac{dV_3}{dt} = M\frac{V_1}{V_{1_{eq}}}\frac{V_2}{V_{2_{eq}}}\left(1 - \frac{V_3}{V_{3_{eq}}}\right). \tag{12.9}$$

12.3.3 Model Results

With $V_1 = V_2 = V_3 = 0$ at $t = 0$, the solutions to Eqs. (12.7), (12.8) and (12.9) are, respectively,

$$V_1 = V_{1_{eq}}\left(1 - e^{-\alpha t}\right), \qquad \text{with } \alpha = \frac{M}{V_{1_{eq}}}, \tag{12.10}$$

$$V_2 = V_{2_{eq}}\left(1 - e^{-\beta t'}\right), \qquad \text{with } \beta = \frac{M}{V_{2_{eq}}}, \qquad \text{and } t' = t - \frac{V_1}{M}, \tag{12.11}$$

$$V_3 = V_{3_{eq}}\left(1 - e^{-\gamma t''}\right), \qquad \text{with } \gamma = \frac{M}{V_{3_{eq}}}, \qquad \text{and } t'' = t' - \frac{V_2}{M}. \tag{12.12}$$

The reader can verify this by substituting the solutions in Eqs. (12.7)–(12.9). The inverses of the coefficients α, β and γ represent timescales. Here, α^{-1} is the adaptation timescale of the ebb tidal shoal, β^{-1} is the adaptation timescale of the bypassing bar in the absence of the ebb tidal shoal and γ^{-1} is the adaptation timescale of the attachment bar in the absence of the ebb tidal shoal and the bypassing bar. Timescales are independent of those of the downdrift elements, implying that there is no feedback.

For Ocean City Inlet, parameter values presented in Kraus (2000) are $M = 0.15 \times 10^6$ m^3 year^{-1}, $V_{1_{eq}} = 3 \times 10^6$ m^3, $V_{2_{eq}} = 7 \times 10^6$ m^3 and $V_{3_{eq}} = 0.5 \times 10^6$ m^3. It is not clear how the values of the equilibrium volumes were determined. The development of the sand volumes of the different elements as calculated from Eqs. (12.10)–(12.12), together with observations, are plotted in Fig. 12.3. The timescale for the ebb tidal shoal is $\alpha^{-1} = 20$ years, for the bypassing bar is $\beta^{-1} = 46$ years and for the attachment bar is $\gamma^{-1} = 3$ years. The model results and

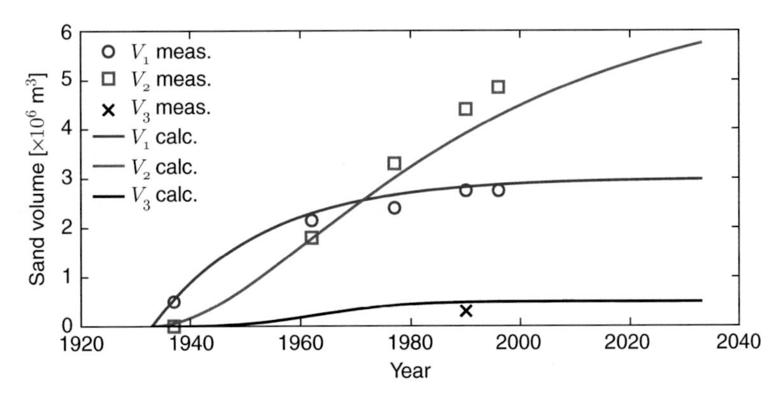

Figure 12.3 Modeled development of ebb tidal shoal, bypassing bar and attachment bar at Ocean City Inlet together with observations; subscript 1 refers to ebb tidal shoal, subscript 2 refers to bypassing bar and subscript 3 refers to attachment bar (adapted from Kraus, 2000, with permission from ASCE).

observations show reasonable agreement. For an in-depth discussion of the results the reader is referred to Kraus (2000).

12.4 Adaptation of the Frisian Inlet after Basin Reduction

12.4.1 Frisian Inlet

The Frisian Inlet is one of the tidal inlets of the Dutch Wadden Sea, located between the island of Schiermonnikoog and the Engelsmanplaat shoal (Fig. 12.4). Major morphological elements are the ebb delta, the inlet (Zoutkamperlaag) and the tidal flats. The water motion in the inlet is governed by the tide, as freshwater inflow is insignificant. Tides are semi-diurnal with an offshore mean tidal range of 2.25 m. The offshore mean annual significant wave height is 1.13 m. The dominant wave direction is from the northwest. Estimates of the gross annual longshore sand transport, M, vary between 0.5×10^5 and 1×10^5 m^3. The sand transport is primarily in a west–east direction.

In 1969, the basin surface area of the inlet was reduced by approximately 30 percent, resulting in a decrease in the tidal prism from 325×10^6 to 225×10^6 m^3. Prior

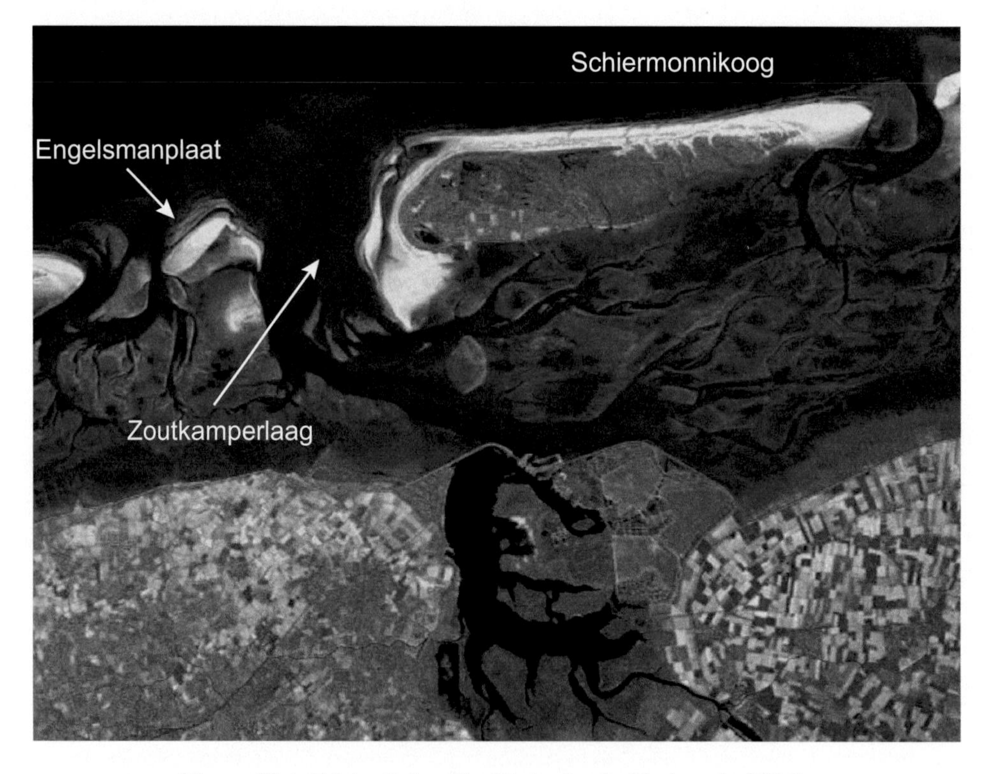

Figure 12.4 Frisian Inlet, The Netherlands (Esri et al., 2016).

to basin reduction, the morphology was in equilibrium with an ebb delta (sand) volume $V_{s_0} = 132 \times 10^6$ m^3 and an inlet (water) volume $V_{w_0} = 171 \times 10^6$ m^3. Following the basin reduction, measurements every four years over an 18-year period showed that by the end of this period the delta volume had decreased by 21×10^6 m^3 and the inlet volume had decreased by 31×10^6 m^3 (Biegel and Hoekstra, 1995). The inlet volume decreased monotonically while the changes in delta volume showed a more random character (see Fig. 12.6 discussed in Section 12.4.3). For example, during the period 1975–1979, instead of a decrease, an increase in delta volume was observed. A possible reason for this random character could be the difficulty in defining the extent of the (moving) delta. During the 18-year period, the tidal prism remained constant at 225×10^6 m^3 and most likely will stay close to this value for the remainder of the adaptation period. Both the observed ebb delta volumes and inlet volumes are used to calibrate the empirical model.

The ebb delta volume and the inlet volume are expected to ultimately reach new equilibrium values. The new equilibrium volume of the ebb delta is estimated using the relationship between ebb delta sand volume and tidal prism for inlets in the Wadden Sea presented in Louters and Gerritsen (1994). With $P = 225 \times 10^6$ m^3, this resulted in an equilibrium sand volume of the delta $V_{s_{eq}} = 83 \times 10^6$ m^3. The equilibrium water volume of the inlet was estimated using the two cross-sectional area – tidal prism relationships for the inlets of the Wadden Sea (Section 5.2.3). Both relationships resulted in an equilibrium cross-sectional area of the gorge of the inlet of approximately $A = 15,300$ m^2. From this, and making certain assumptions for the variation in cross-sectional area over the inlet, the equilibrium water volume of the inlet is estimated at $V_{w_{eq}} = 118 \times 10^6$ m^3 (van de Kreeke, 2006). After the basin reduction, the ebb delta has a surplus and the inlet has a shortage of sand. When the new equilibrium is reached, the delta and inlet combined have gained a volume of sand equal to 4×10^6 m^3. This gain comes at the expense of the downdrift coast.

12.4.2 Schematization and Model Formulation

To model the adaptation of the morphology after basin reduction, the tidal inlet is represented by two of the three major elements, the ebb delta and the inlet. The third element, the tidal flats, is not included as its response time is much larger than that of the ebb delta and inlet (Biegel and Hoekstra, 1995). As shown in Fig. 12.5, elements are connected in series. Contrary to the model for Ocean City Inlet there is feedback, in this case between the ebb delta and the inlet. The ebb delta is characterized by the sand volume V_s and the inlet by the water volume V_w. A fraction M_1 of the longshore sand transport enters the delta and the remaining fraction M_2 enters the inlet. S_1 is the sand transport from the inlet to the delta resulting from

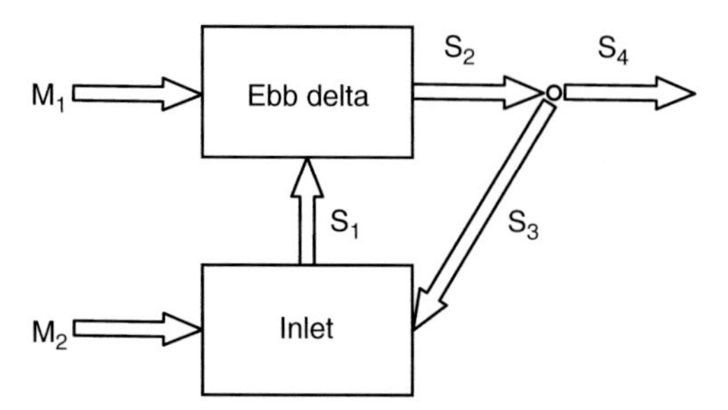

Figure 12.5 Empirical model for the Frisian Inlet; for definition of the symbols reference is made to the text.

the ebb tidal current. S_2 is the wave-induced sand transport leaving the delta in the direction of the downdrift coast. Before reaching the downdrift coast, a part of this sand volume, S_3, is diverted to the inlet. The fraction reaching the downdrift coast is $S_4 = S_2 - S_3$. Because the interest is in timescales ranging from decades to centuries, sand transports are taken as annually averaged transports.

The sand transport from the inlet to the delta is taken as being proportional to the ratio of the equilibrium and the actual water volume of the inlet to a power n:

$$S_1 = \kappa_1 \left(\frac{V_{w_{eq}}}{V_w} \right)^n, \tag{12.13}$$

where $V_{w_{eq}}$ is the equilibrium water volume of the inlet, n is an empirical constant and κ_1 is a proportionality constant. The sand transport from the delta towards the downdrift coast is taken as being proportional to the ratio of the actual and equilibrium sand volume of the delta to a power m:

$$S_2 = \kappa_2 \left(\frac{V_s}{V_{s_{eq}}} \right)^m, \tag{12.14}$$

where $V_{s_{eq}}$ is the equilibrium sand volume of the ebb delta, m is an empirical constant and κ_2 is a proportionality constant. A difference with the reservoir model (Section 12.3) is that the sand transport leaving an element is solely dependent on the morphological state of that element and is independent of the sand transport entering the element.

The transport diverted from the delta to the inlet, S_3, is taken as a fraction α of S_2, i.e.,

$$S_3 = \alpha S_2, \tag{12.15}$$

with $0 \leq \alpha \leq 1$. Through this feedback mechanism, the adaptation of the inlet depends on that of the ebb delta.

The transport to the downdrift coast is

$$S_4 = (1 - \alpha)S_2. \tag{12.16}$$

When at equilibrium, the sand transport leaving equals the sand transport entering an element. For the inlet element,

$$M_2 - S_{1_{eq}} + \alpha S_{2_{eq}} = 0, \tag{12.17}$$

and for the delta element

$$M_1 + S_{1_{eq}} - S_{2_{eq}} = 0, \tag{12.18}$$

with the subscript *eq* referring to equilibrium conditions. Solving for $S_{1_{eq}}$ and $S_{2_{eq}}$ from Eqs. (12.17) and (12.18) results in

$$S_{1_{eq}} = \frac{\alpha M_1 + M_2}{1 - \alpha}, \tag{12.19}$$

and

$$S_{2_{eq}} = \frac{M_1 + M_2}{1 - \alpha}. \tag{12.20}$$

Using Eq. (12.13), it follows that $S_{1_{eq}} = \kappa_1$ and using Eq. (12.14), $S_{2_{eq}} = \kappa_2$. It then follows from Eqs. (12.19) and (12.20) that

$$\kappa_1 = \frac{\alpha M_1 + M_2}{1 - \alpha}, \tag{12.21}$$

and

$$\kappa_2 = \frac{M_1 + M_2}{1 - \alpha}. \tag{12.22}$$

Substituting for $\kappa_1$from Eq. (12.21) in Eq. (12.13) results in

$$S_1 = \frac{\alpha M_1 + M_2}{1 - \alpha}\left(\frac{V_{w_{eq}}}{V_w}\right)^n. \tag{12.23}$$

Substituting for κ_2 from Eq. (12.22) in Eq. (12.14) results in

$$S_2 = \frac{M_1 + M_2}{1 - \alpha}\left(\frac{V_s}{V_{s_{eq}}}\right)^m. \tag{12.24}$$

Using the expressions for the transports S_1, S_2 and S_3 and assuming that excess volumes of sand entering an element are deposited on the bed and excess volumes of sand leaving an element are eroded from the bed, the conservation of sand equation for the delta is

$$\frac{dV_s}{dt} + \frac{M_1 + M_2}{1 - \alpha}\left(\frac{V_s}{V_{s_{eq}}}\right)^m - \frac{\alpha M_1 + M_2}{1 - \alpha}\left(\frac{V_{w_{eq}}}{V_w}\right)^n = M_1, \tag{12.25}$$

and for the inlet is

$$\frac{dV_w}{dt} + \frac{\alpha(M_1 + M_2)}{1 - \alpha}\left(\frac{V_s}{V_{s_{eq}}}\right)^m - \frac{\alpha M_1 + M_2}{1 - \alpha}\left(\frac{V_{w_{eq}}}{V_w}\right)^n = -M_2. \qquad (12.26)$$

The two coupled differential equations (12.25) and (12.26) describe the transition of the delta and the inlet from the old to the new equilibrium.

12.4.3 Model Results

Eqs. (12.25) and (12.26) are solved numerically for V_s and V_w with the observed initial conditions $V_{w0} = 171 \times 10^6$ m^3 and $V_{s0} = 132 \times 10^6$ m^3 and the equilibrium sand and water volumes are $V_{s_{eq}} = 83 \times 10^6$ m^3 and $V_{w_{eq}} = 118 \times 10^6$ m^3 (Section 12.4.1). Values of the parameters M_1, M_2, m, n and α are determined by matching calculated and observed values of the inlet water volume V_w and the delta sand volumes V_s. In view of the aberrant behavior of the delta volume around year 10, emphasis in the calibration has been on matching observed and calculated inlet volumes. With the tidal prism $P = 225 \times 10^6$ m^3 and the longshore sand transport $M < 1 \times 10^6$ m^3 year^{-1} the ratio $P/M > 225$. Referring to Section 3.3.2, it follows that for the Frisian Inlet the sand bypassing mode is tidal flow bypassing. Therefore, the fraction of the longshore sand transport M_2 entering the inlet is assumed to be larger than the fraction of the longshore sand transport M_1 entering the delta. With the constraints that 0.5×10^6 m^3 year^{-1} $< M < 1 \times 10^6$ m^3 year^{-1} and $0 \leq \alpha < 1$, a good match between observed and calculated values of V_s and V_w during the first 18 years after basin reduction was found for $M_1 = 0.25 \times 10^6$ m^3 year^{-1}, $M_2 = 0.75 \times 10^6$ m^3 year^{-1}, $m = 0.9$, $n = 3$ and $\alpha = 0.65$; see Figs. 12.6a and 12.6b.

Using Eqs. (12.25) and (12.26) and the parameter values determined by matching observed and calculated values of V_s and V_w for the 18-year observational period, the volume deficits $V_s - V_{s_{eq}}$ and $V_w - V_{w_{eq}}$ are calculated for a period of 200 years. The results are plotted in Fig. 12.7. The trajectory towards equilibrium for the inlet shows a monotonic decrease in the water volume. Initially, the delta volume decreases but then shows an overshoot before approaching equilibrium. This is further discussed in Section 12.4.5.

For the same 200-year period, sand transport values are presented in Fig. 12.8. The transport from inlet to delta (S_1) increases monotonically to an equilibrium value of $S_{1_{eq}} = 2.61 \times 10^6$ m^3 year^{-1}. The transport from the delta (S_2) shows a slight overshoot; it decreases to reach a minimum approximately 60 years after basin reduction and then increases to the equilibrium value of $S_{2_{eq}} = 2.86 \times 10^6$ m^3 year^{-1}. Because S_3 and S_4 are directly related to S_2, they both show a similar overshoot.

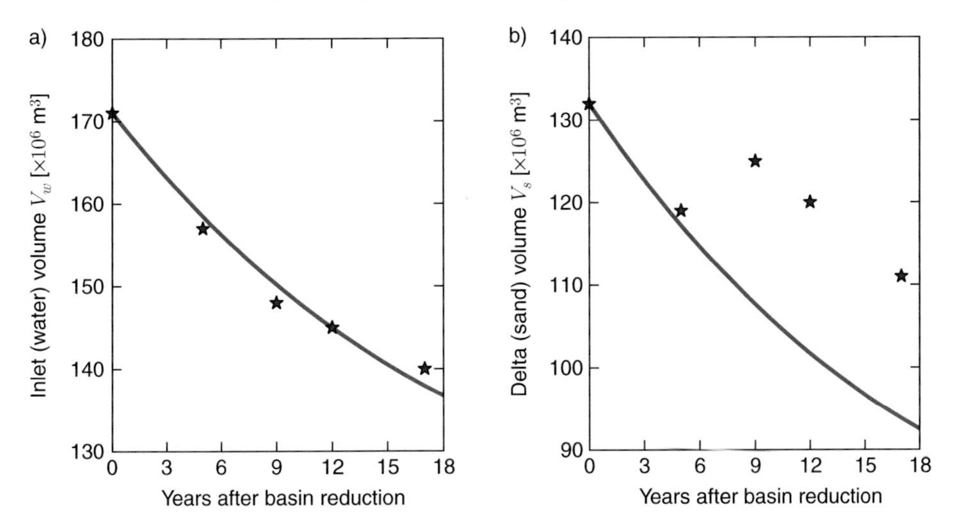

Figure 12.6 Observed (star) and numerically calculated (solid red line) values of a) the water volume of the inlet V_w and b) the sand volume of the delta V_s.

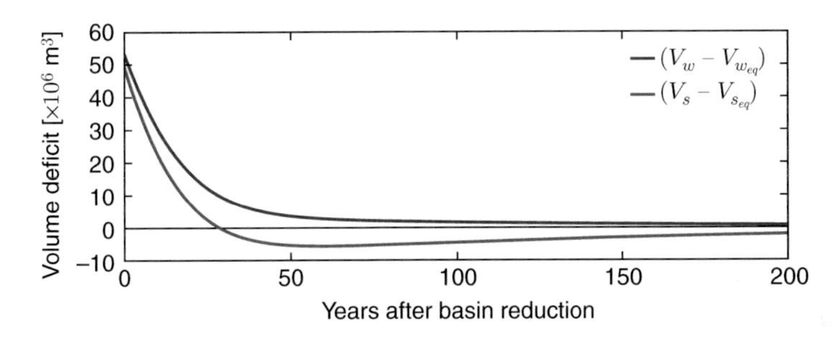

Figure 12.7 Changes in delta (sand) volume deficit and inlet (water) volume deficit for a 30 percent basin reduction using the numerical solution to Eqs. (12.25) and (12.26); for definition of the symbols reference is made to the text.

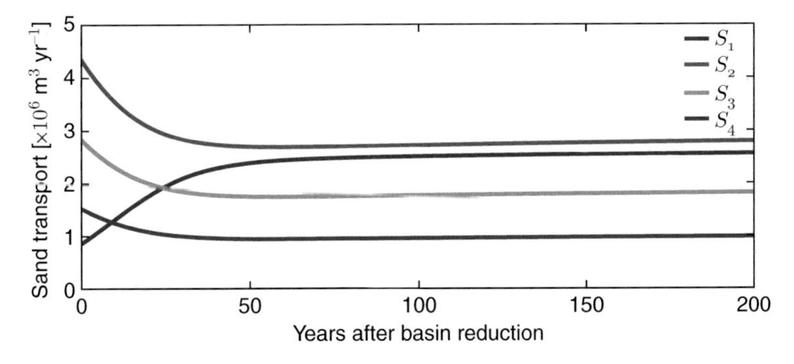

Figure 12.8 Sand transport after basin reduction; for definition of the symbols reference is made to the text.

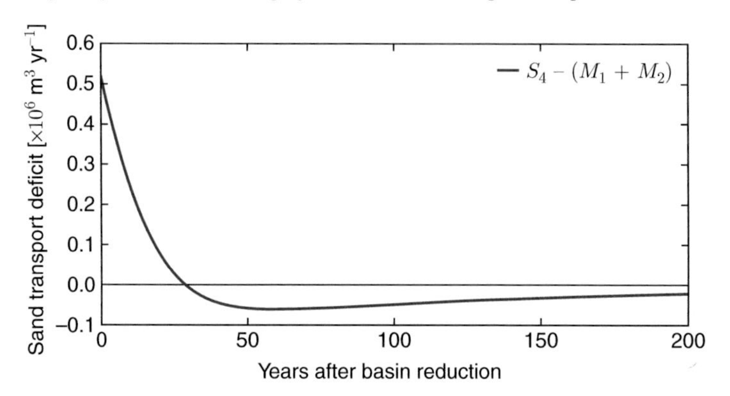

Figure 12.9 Difference between the sand transport to the downdrift coast and the longshore sand transport; for definition of the symbols reference is made to the text.

Because of its effect on the downdrift beaches, the transport S_4 is of particular interest. Referring to Fig. 12.9, initially this transport is considerably larger than the longshore sand transport of 1.0×10^6 m^3 year^{-1}, potentially resulting in accretion of the downdrift beaches. After approximately 40 years the value of S_4 becomes, less than the longshore sand transport and at that time erosion of the downdrift beaches should be expected.

12.4.4 Analytical Solution; Local and System Timescales

To explain the numerically calculated evolution of the delta and inlet and to identify the relevant timescales, an approximate analytical solution to Eqs. (12.25) and (12.26) is presented. For this, the quotients $(V_s/V_{s_{eq}})^m$ and $(V_{w_{eq}}/V_w)^n$ are linearized by imposing the condition that the morphological state is close to equilibrium, i.e.,

$$\frac{|V_{s_o} - V_{s_{eq}}|}{V_{s_{eq}}} \ll 1, \qquad \text{and} \qquad \frac{|V_{w0} - V_{w_{eq}}|}{V_{w_{eq}}} \ll 1. \tag{12.27}$$

In that case, to a good approximation

$$\left(\frac{V_s}{V_{s_{eq}}}\right)^m = 1 + m\left(\frac{V_s - V_{s_{eq}}}{V_{s_{eq}}}\right), \tag{12.28}$$

and

$$\left(\frac{V_{w_{eq}}}{V_w}\right)^n = 1 - n\left(\frac{V_w - V_{w_{eq}}}{V_{w_{eq}}}\right). \tag{12.29}$$

For a detailed description of the method used to derive Eqs. (12.28) and (12.29) reference is made to Section 8.3.

To satisfy the conditions expressed by Eq. (12.27), a hypothetical case is considered in which the basin surface area of the Frisian Inlet is reduced by slightly less than 10 percent, resulting in a reduction in tidal prism from 325×10^6 m^3 to 300×10^6 m^3. The water volume of the inlet and the sand volume of the delta before basin reduction are $V_{w_0} = 171 \times 10^6$ m^3 and $V_{s_0} = 132 \times 10^6$ m^3, respectively. Using the cross-sectional area – tidal prism relationship for the inlets of the Dutch Wadden Sea (Section 5.2.3), it follows that with a tidal prism of 300×10^6 m^3, the equilibrium inlet cross-sectional area after basin reduction is 20,400 m^2. Making certain assumptions for the variation in cross-sectional area over the inlet, the corresponding water volume of the inlet $V_{w_{eq}} = 158 \times 10^6$ m^3 (van de Kreeke, 2006). The equilibrium sand volume of the ebb delta follows from the relationship between ebb delta sand volume and tidal prism for tidal inlets in the Wadden Sea (Louters and Gerritsen, 1994). With the tidal prism of 300×10^6 m^3, this result in an equilibrium sand volume is $V_{s_{eq}} = 119 \times 10^6$ m^3. Substitution of the different sand and water volumes in Eq. (12.27) shows that the condition imposed by this equation is satisfied.

Starting with Eqs. (12.28) and (12.29) and introducing the timescales,

$$\tau_s = \frac{V_{s_{eq}}(1-\alpha)}{m(M_1+M_2)}, \quad \text{and} \quad \tau_w = \frac{V_{w_{eq}}(1-\alpha)}{n(\alpha M_1+M_2)}. \tag{12.30}$$

Eqs. (12.25) and (12.26) are written as, respectively,

$$\frac{d(V_s - V_{s_{eq}})}{dt} + \frac{(V_s - V_{s_{eq}})}{\tau_s} + \frac{(V_w - V_{w_{eq}})}{\tau_w} = 0, \tag{12.31}$$

and

$$\frac{d(V_w - V_{w_{eq}})}{dt} + \frac{(V_w - V_{w_{eq}})}{\tau_w} + \alpha\frac{(V_s - V_{s_{eq}})}{\tau_s} = 0. \tag{12.32}$$

It follows from Eq. (12.31) that when the inlet is in equilibrium, the delta adapts exponentially with the timescale τ_s. Similarly, when the delta is in equilibrium, it follows from Eq. (12.32) that the inlet adapts exponentially with the timescale τ_w. As these timescales do not involve the interaction of the two elements, they are referred to as local timescales.

Defining $V_s - V_{s_{eq}} = y_1$ and $V_w - V_{w_{eq}} = y_2$, Eqs. (12.31) and (12.32) in matrix form are

$$\begin{bmatrix} \frac{dy_1}{dt} \\ \frac{dy_2}{dt} \end{bmatrix} = \begin{bmatrix} -\frac{1}{\tau_s} & -\frac{1}{\tau_w} \\ -\frac{\alpha}{\tau_s} & -\frac{1}{\tau_w} \end{bmatrix} \begin{bmatrix} y_1 \\ y_2 \end{bmatrix}. \tag{12.33}$$

The solution to Eq. (12.33) is

$$\begin{pmatrix} y_1 \\ y_2 \end{pmatrix} = C_1 \begin{pmatrix} 1 \\ x_1 \end{pmatrix} e^{\lambda_1 t} + C_2 \begin{pmatrix} 1 \\ x_2 \end{pmatrix} e^{\lambda_2 t}, \tag{12.34}$$

in which λ_1 and λ_2 are the eigenvalues of the square matrix and $\binom{1}{x_1}$ and $\binom{1}{x_2}$ are the corresponding eigenvectors. The coefficients C_1 and C_2 are determined by the initial conditions. In terms of the timescales τ_s and τ_w and the feedback coefficient α, the eigenvalues are

$$\lambda_1 = \frac{-\left(\frac{1}{\tau_s}+\frac{1}{\tau_w}\right)+\sqrt{\left(\frac{1}{\tau_s}+\frac{1}{\tau_w}\right)^2+\frac{4(\alpha-1)}{\tau_s\tau_w}}}{2}, \quad (12.35)$$

$$\lambda_2 = \frac{-\left(\frac{1}{\tau_s}+\frac{1}{\tau_w}\right)-\sqrt{\left(\frac{1}{\tau_s}+\frac{1}{\tau_w}\right)^2+\frac{4(\alpha-1)}{\tau_s\tau_w}}}{2}. \quad (12.36)$$

With $0 \leq \alpha < 1$, both λ_1 and λ_2 are negative with $|\lambda_1| < |\lambda_2|$. The reciprocals of the absolute value of the eigenvalues, $\tau_1 = 1/|\lambda_1|$ and $\tau_2 = 1/|\lambda_2|$ are referred to as the system timescales. As opposed to the local timescales, τ_s and τ_w, the system timescales account for the interaction of the two elements. For additional interpretation of the system timescales, reference is made to the last paragraph of this section. In terms of the system timescales, the solution for y_1 and y_2 is

$$\begin{pmatrix} y_1 \\ y_2 \end{pmatrix} = C_1 \begin{pmatrix} 1 \\ x_1 \end{pmatrix} e^{-\frac{t}{\tau_1}} + C_2 \begin{pmatrix} 1 \\ x_2 \end{pmatrix} e^{-\frac{t}{\tau_2}}. \quad (12.37)$$

The components of the eigenvectors are

$$x_1 = -\tau_w\left(\frac{1}{\tau_s}+\lambda_1\right), \quad \text{and} \quad x_2 = -\tau_w\left(\frac{1}{\tau_s}+\lambda_2\right). \quad (12.38)$$

Substituting for λ_1 and λ_2 from, respectively, Eqs. (12.35) and (12.36) in Eq. (12.38) results in the expressions for the components of the eigenvectors

$$x_{1,2} = \frac{-\left(\frac{1}{\tau_s}-\frac{1}{\tau_w}\right)\mp\sqrt{\left(\frac{1}{\tau_s}+\frac{1}{\tau_w}\right)^2-4(1-\alpha)\left(\frac{1}{\tau_s}\frac{1}{\tau_w}\right)}}{\frac{2}{\tau_w}}, \quad (12.39)$$

with the negative sign referring to x_1 and the positive sign referring to x_2. This equation can be written as

$$x_{1,2} = \frac{-\left(\frac{1}{\tau_s}-\frac{1}{\tau_w}\right)\mp\sqrt{\left(\frac{1}{\tau_s}-\frac{1}{\tau_w}\right)^2+4\alpha\left(\frac{1}{\tau_s}\frac{1}{\tau_w}\right)}}{\frac{2}{\tau_w}}. \quad (12.40)$$

Because $0 \leq \alpha < 1$, the square root in Eq. (12.40) is always larger than $|1/\tau_s - 1/\tau_w|$. It follows that the components of the eigenvectors, x_1 and x_2, have opposite signs.

With the initial conditions for the sand volume of the delta, $V_s = V_{s0}$ and for the water volume of the inlet, $V_w = V_{w0}$, the expressions for C_1 and C_2 are

$$C_1 = \frac{(V_{s0} - V_{s_{eq}})x_2 - (V_{w0} - V_{w_{eq}})}{x_2 - x_1}, \tag{12.41}$$

and

$$C_2 = \frac{-(V_{s0} - V_{s_{eq}})x_1 + (V_{w0} - V_{w_{eq}})}{x_2 - x_1}. \tag{12.42}$$

To illustrate the adaptation of the delta volume and the inlet volume, the parameter values for M_1, M_2, m, n and α in Section 12.4.3 for the basin reduction of 30 percent are assumed to also be valid for the basin reduction of 10 percent. With these values and $V_{s_{eq}} = 119 \times 10^6$ m^3 and $V_{w_{eq}} = 158 \times 10^6$ m^3, the values of the local timescales are $\tau_s = 46$ years and $\tau_w = 20$ years. The eigenvalues are $\lambda_1 = -0.0056$ year^{-1} and $\lambda_2 = -0.0656$ year^{-1}, resulting in the system timescales $\tau_1 = 1/|\lambda_1| = 118$ years and $\tau_2 = 1/|\lambda_2| = 15$ years. Values of the eigenvector are $x_1 = -0.38$ and $x_2 = 0.83$. With the initial deficits, $V_{s0} - V_{s_{eq}} = 13 \times 10^6$ m^3 and $V_{w0} - V_{w_{eq}} = 13 \times 10^6$ m^3, the values of the coefficients C_1 and C_2 are -1.9×10^6 m^3 and 14.9×10^6 m^3, respectively. Using the results of the analytical solution, the adaptation of the delta volume and inlet volume are presented in Fig. 12.10. Comparing with Fig. 12.7, it follows that the analytically calculated adaptation for the 10 percent basin reduction is qualitatively the same as the numerically calculated adaptation for the 30 percent basin reduction. The inlet (water) volume shows a monotonic decrease until it reaches the equilibrium value. Initially, the delta (sand) volume decreases but then overshoots before returning to the equilibrium value.

Referring to Eq. (12.37), the fractions of the initial sand and water volume deficits adjusting with the shorter system timescale τ_2 are C_2 and C_2x_2, respectively. With C_2 and x_2 positive, both volumes have the same sign, i.e., the sand volume of the delta and the water volume of the inlet are both too large. This

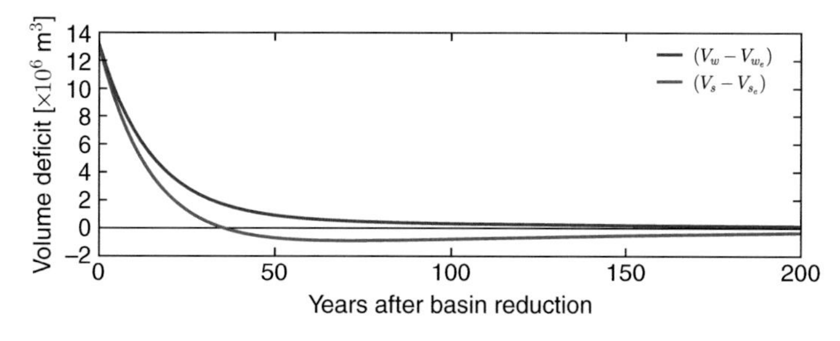

Figure 12.10 Changes in delta (sand) volume deficit and inlet (water) volume deficit for a 10 percent basin reduction using the analytical solution, Eq. (12.37).

results in an internal distribution of sand, whereby the shortage of sand in the inlet is partially compensated by a supply of sand from the delta. The fractions of the initial sand and water volume deficits adjusting with the longer timescale, τ_1, are C_1 and C_1x_1, respectively. With C_1 and x_1 negative, the sand volume and water volume have opposite signs, i.e., the sand volume of the delta is too small and the water volume of the inlet is too large. Compensation of the shortage of sand in both the delta and inlet has to come from outside, which explains the longer timescale.

12.4.5 Bumps and Overshoots

With the eigenvalues λ_1 and λ_2 negative, both the delta and the inlet evolve towards an equilibrium state. As explained in Kragtwijk et al. (2004), this evolution is not necessarily monotonic. Depending on the interaction of the two elements, the initial response of an element may be away from its equilibrium, referred to as bump behavior. As shown in Figs. 12.7 and 12.10, it is also possible for an element to overshoot its equilibrium. These two contrasting situations are illustrated for the delta element in Fig. 12.11. In the case of a bump, even though there is already a surplus, sand is added to the delta. In the case of an overshoot, even though the delta element has reached equilibrium, additional sand is withdrawn.

The criterion for a bump is that the sign of the initial rate of change of the volume deficit and the sign of the initial volume deficit are the same. For the 10 percent basin reduction, the terms in Eq. (12.37) with the shortest timescale are positive. As a result, the rates of change of the initial inlet and delta volumes are negative. Because the initial volume deficits for both inlet and delta are positive, neither the evolution of the delta nor that of the inlet shows a bump. This agrees with the results in Figs. 12.7 and 12.10.

The criterion for an overshoot is that the initial volume deficit and the volume deficit close to equilibrium have different signs. For the 10 percent basin reduction, both the initial delta and inlet volume deficits are positive. Referring to the terms

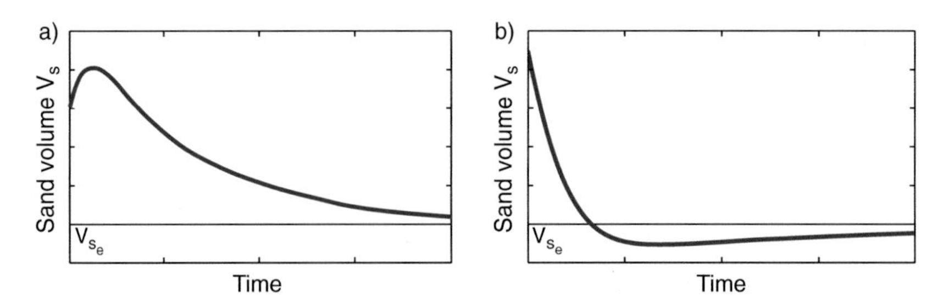

Figure 12.11 Schematic of a bump a) and an overshoot b) for a delta element. V_s is sand volume of the delta and V_{s_e} is equilibrium sand volume of the delta.

with the longest timescale in Eq. (12.37), it follows that the volume deficit for the delta is negative and for the inlet is positive. The evolution of the delta sand volume will show an overshoot and the evolution of the inlet water volume will be monotonic. This agrees with the results in Figs. 12.7 and 12.10.

12.5 Adaptation of an Inlet-Delta System Using a Diffusive Transport Formulation

A modeling approach, using a diffusive transport between the elements, was introduced by Di Silvio (1989) and later expanded by Stive et al. (1998). The principles are illustrated using a schematized inlet-ebb delta system (Fig. 12.12). Similar to the examples presented in Sections 12.3 and 12.4, it is assumed that after a perturbation the inlet and delta return to an equilibrium state. In Fig. 12.12, Element 1 represents the inlet with a water volume V_w and Element 2 represents the delta with a sand volume V_s. Sand enters and leaves Element 1 through exchange with Element 2. Sand enters and leaves Element 2 through exchange with Element 1 and the outside world (ocean). The diffusive sand transport between Elements 1 and 2 is

$$S_{12} = \delta(c_1 - c_2). \tag{12.43}$$

Similarly, the transport between Element 2 and the outside world is

$$S_{20} = \delta(c_2 - c_E). \tag{12.44}$$

In these equations, c_1 and c_2 are the local volume concentrations of suspended sand in Element 1 and 2, respectively, c_E is the volume concentration of suspended sand in the outside world and δ is a diffusion coefficient. The concentration c_E is assumed to remain constant during the transition period. To simplify the algebra, the diffusion coefficient δ is taken the same for Element 1 and 2. Transports are annually averaged values. The diffusive type transport results in a feedback between the ebb delta and the inlet.

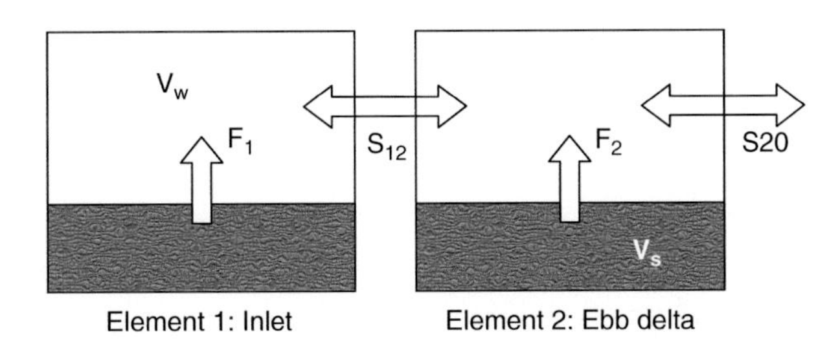

Figure 12.12 Two-element system with diffusive transport; for definition of the symbols see text.

The volume flux of sand from the bottom to the water column in Element 1 is

$$F_1 = w_s A(c_{1_e} - c_1), \tag{12.45}$$

in which c_{1_e} is the local equilibrium sand concentration in the water column. The volume flux of sand from the bottom to the water column in Element 2 is

$$F_2 = w_s A(c_{2_e} - c_2), \tag{12.46}$$

in which c_{2_e} is the local equilibrium sand concentration in the water column. In Eqs. (12.45) and (12.46), w_s is an exchange coefficient and A is the bottom surface area. To simplify the algebra, the bottom surface areas are taken the same for both elements.

The conservation of suspended sand equation for Element 1 is

$$\frac{dV_{sus}}{dt} = F_1 - S_{12} = 0, \tag{12.47}$$

and for Element 2 is

$$\frac{dV_{sus}}{dt} = F_2 + S_{12} - S_{20} = 0. \tag{12.48}$$

In these equations, V_{sus} is the suspended sand volume in the element. Assuming that on an annual average base the volume of suspended sand in the water column remains constant, $dV_{sus}/dt = 0$. It then follows from Eq. (12.47), with S_{12} given by Eq. (12.43) and F_1 given by (12.45), that

$$w_s A(c_{1_e} - c_1) - \delta(c_1 - c_2) = 0. \tag{12.49}$$

Similarly, it follows from Eq. (12.48), with S_{12} given by Eq. (12.43), S_{20} given by Eq. (12.44) and F_2 given by (12.46), that

$$w_s A(c_{2_e} - c_2) + \delta(c_1 - c_2) - \delta(c_2 - c_E) = 0. \tag{12.50}$$

Solving for c_1 and c_2 from Eqs. (12.49) and (12.50) results in

$$c_1 = \frac{w_s A(w_s A + 2\delta)c_{1_e} + w_s A\delta c_{2_e} + \delta^2 c_E}{(w_s A + 2\delta)(w_s A + \delta) - \delta^2}, \tag{12.51}$$

and

$$c_2 = \frac{w_s A\delta c_{1_e} + w_s A(w_s A + \delta)c_{2_e} + \delta(w_s A + \delta)c_E}{(w_s A + 2\delta)(w_s A + \delta) - \delta^2}. \tag{12.52}$$

For the inlet, the local and the overall equilibrium concentrations are related by the ratio of the equilibrium water volume and the actual water volume. For the delta, these concentrations are related by the ratio of the actual sand volume and the equilibrium sand volume (van Goor, 2003), i.e.,

$$c_{1_e} = c_E \left(\frac{V_{w_e}}{V_w}\right)^n, \tag{12.53}$$

$$c_{2_e} = c_E \left(\frac{V_s}{V_{s_e}} \right)^m . \tag{12.54}$$

In these equations, V_w is the actual water volume and V_{w_e} is equilibrium water volume of Element 1. V_s is the actual sand volume and V_{s_e} is the equilibrium sand volume of Element 2. The exponents n and m are empirical coefficients. When the system is at equilibrium, $c_1 = c_{1_e} = c_E$ and $c_2 = c_{2_e} = c_E$.

The rate of change of the water volume V_w and the sand volume V_s are related to the volume flux of sand from the bottom to the water column by, respectively,

$$\frac{dV_s}{dt} = -F_2 = -w_s A(c_{2_e} - c_2), \tag{12.55}$$

and

$$\frac{dV_w}{dt} = F_1 = w_s A(c_{1_e} - c_1). \tag{12.56}$$

Using Eqs. (12.51)–(12.54), it follows from Eq. (12.55) that

$$\frac{dV_s}{dt} + (\mu_1 + 2\mu_2) c_E \left(\frac{V_s}{V_{s_e}} \right)^m - \mu_2 c_E \left(\frac{V_{w_e}}{V_w} \right)^n = (\mu_1 + \mu_2) c_E, \tag{12.57}$$

with the coefficients μ_1 and μ_2 defined as

$$\mu_1 = \frac{w_s A \delta^2}{((w_s A + 2\delta)(w_s A + \delta) - \delta^2)}, \tag{12.58}$$

and

$$\mu_2 = \frac{(w_s A)^2 \delta}{((w_s A + 2\delta)(w_s A + \delta) - \delta^2)}. \tag{12.59}$$

Using Eqs. (12.51)–(12.54), it follows from Eq. (12.56) that

$$\frac{dV_w}{dt} + \mu_2 c_E \left(\frac{V_s}{V_{s_e}} \right)^m - (\mu_1 + \mu_2) c_E \left(\frac{V_{w_e}}{V_w} \right)^n = -\mu_1 c_E. \tag{12.60}$$

Similar to Eqs.(12.25) and (12.26), the nature of the solution to Eqs. (12.57) and (12.60) can be investigated by assuming a small perturbation. This allows linearizing the nonlinear terms in these equations, resulting in

$$\frac{d(V_s - V_{s_e})}{dt} + (\mu_1 + 2\mu_2) \frac{m c_E}{V_{s_e}} (V_s - V_{s_e}) + \mu_2 \frac{n c_E}{V_{w_e}} (V_w - V_{w_e}) = 0, \tag{12.61}$$

and

$$\frac{d(V_w - V_{w_e})}{dt} + \mu_2 \frac{m c_E}{V_{s_e}} (V_s - V_{s_e}) + (\mu_1 + \mu_2) \frac{n c_E}{V_{w_e}} (V_w - V_{w_e}) = 0. \tag{12.62}$$

Except for the coefficients, these equations are the same as Eqs. (12.31) and (12.32), respectively. In view of this, the response of the inlet-ebb delta system to a perturbation using a diffusive transport formulation is similar to the response using the model described in Section 12.4.4.

Applications of the empirical model using diffusive transport are found in van Goor (2003) and Kragtwijk et al. (2004). van Goor (2003) deals with the impact of sea level rise on the morphological equilibrium state of tidal inlets, with application to two inlets in the Dutch Wadden Sea: the Ameland Inlet described in Section 4.8 and the Eyerlandse Gat Inlet. In Kragtwijk et al. (2004) the focus is on the evolution of the Dutch Wadden Sea inlets after closure of the Zuiderzee. In this study, special attention is given to the Texel and Vlie Inlets, described in Chapter 9, as they are most affected by the closure.

12.6 Limitations of Empirical Modeling

Empirical modeling of tidal inlets is directed at the transition of the morphology from a perturbed to an equilibrium state. A requirement is that the equilibrium state of the elements of the tidal inlet is known. The equilibrium state is expressed in terms of a relationship between the sand or water volume of the element and the tidal prism. The formulation of the exchange of sediment between the various morphological elements is empirical and involves a large number of parameters. Many of these lack a physical base. As a result, models need extensive calibration, requiring data over a relatively long period of time. Because of the lack of a physical base, parameter values found for one tidal inlet do not necessarily apply to another tidal inlet (Wang et al., 2008).

The assumption of the inlet system tending towards an equilibrium is not always valid. An example is the reduction of the cross-sectional area of an inlet below a certain value where, instead of tending towards an equilibrium, the inlet closes.

Provided they are properly calibrated over a sufficient long time period, empirical models can provide valuable information over the transition from a perturbed to an equilibrium state and the timescales involved. They are a useful addition to process-based morphodynamic models until some of the shortcomings in these models, preventing application to problems with timescales of decades and centuries, have been resolved.

13

River Flow and Entrance Stability

13.1 Introduction

In the preceding chapters, the emphasis is on tidal inlets where tidal currents are dominant and river flow is of secondary importance. Sand is carried towards the inlets by longshore sand transport and cross-shore sand transport is small. Inlets are open at all times. The few that closed did so gradually through spit formation, thereby increasing the inlet length and decreasing the inlet velocity. Examples are Captain Sam's Inlet and Mason Inlet, both located in South Carolina and described, respectively, in Sections 4.4 and 4.5.

A different category of inlets is where river flow is dominant and the tide is of secondary importance in keeping the inlet open. Inlets in this category are found in Vietnam (Lam, 2009; Tung, 2011), South Africa (Cooper, 2001; Whitfield, 1992) and Australia (Baldock et al., 2008; Hinwood and McLean, 2015b; Hinwood et al., 2012; Morris and Turner, 2010; Ranasinghe and Pattiaratchi, 2003). Many of these inlets connect to small lagoons and have a small tidal range, resulting in a small tidal prism. In addition to longshore sand transport, cross-shore sand transport plays an important role in carrying the sand towards the entrance. The river flow shows strong inter-annual variations with periods of high alternating with periods of low river flow. The height and the period between peak flows have an important bearing on whether the inlets stay open or close. In this respect, a distinction is made between inlets that remain open at all times and inlets that are open only seasonally or intermittently.

Even though only open part of the year, many seasonally and intermittently closed inlets are used extensively as harbors for small fishing boats and as recreational areas for swimming and boating. Closure presents a three-fold problem. Firstly, ocean access for boats that use the back-barrier lagoon as a harbor is limited to when the inlet is open. Secondly, the water quality in the lagoon could deteriorate during the months of inlet closure. Thirdly, flooding of low-lying land may disrupt land use and access and lead to land siltation. Consequently, there is an interest in keeping the inlet permanently open.

In the following sections, the effect of river flow on entrance stability is discussed for three inlets, Thuan An Inlet on the central coast of Vietnam, Wilson Inlet in southwestern Australia and Lake Conjola Inlet on the southeast coast of Australia. Thuan An is an inlet that, in spite of the seasonal character of the river flow, remains open at all times. The seasonal variations in river flow at Wilson Inlet cause this inlet to be open only part of the year. The river flow at Lake Conjola Inlet is highly irregular, resulting in an intermittently open entrance.

Prior to dealing with the three selected inlets, the effect of river flow on the basin tide and inlet velocity is discussed. For this, the Öszoy–Mehta Solution (Section 6.4) is expanded to include river flow.

13.2 Effect of River Flow on Basin Tide and Inlet Velocity

River flow results in a basin water level set-up, a net ebb velocity and a lowering of the amplitudes of the basin tide and the inlet velocity. The mean inlet velocity resulting from variations of depth with tidal stage discussed in (Section 7.4.2) is usually small compared to the mean velocity generated by the river flow.

The effect of the river flow on basin tide and inlet velocity is demonstrated using an expanded version of a lumped parameter model that includes river flow. The dynamic equation for the expanded lumped parameter model is the same as for the lumped parameter model without river flow and is given by Eq. (6.7), repeated here as Eq. (13.1):

$$K_2^2 \frac{du^*}{dt^*} + \frac{1}{K_1^2} u^* |u^*| = \eta_0^* - \eta_b^*. \tag{13.1}$$

In this equation, density gradients resulting from the river flow are neglected. The variables u^*, η_b^*, η_0^* and t^* are the non-dimensional velocity, basin water level, ocean tide and time, respectively, defined by Eq. (6.5). Velocities are positive in the flood direction. The coefficients K_1 and K_2 are defined by Eq. (6.9). Continuity is expressed by Eq. (6.8) with a term added to account for river flow, resulting in

$$u^* = \frac{d\eta_b^*}{dt^*} - Q^*, \tag{13.2}$$

with the non-dimensional river discharge Q^* defined as

$$Q^* = \frac{Q}{\sigma \hat{\eta}_0 A_b}. \tag{13.3}$$

In this equation, Q is the river discharge, $\sigma\hat{\eta}_0 A_b$ is the tidal discharge scale and A_b is the basin surface area. Assumed is a sinusoidal ocean tide with frequency σ and amplitude $\hat{\eta}_0$.

A semi-analytical solution to Eqs. (13.1) and (13.2) forced by a sinusoidal ocean tide is presented in Appendix 13.A and Escoffier and Walton (1979). The solution presented in Appendix 13.A is an expanded version of the Öszoy–Mehta Solution presented in Section 6.4. In both solutions the basin water level and inlet velocity consist of a time-dependent and a time-independent part. For the basin water level,

$$\eta_b^* = \tilde{\eta}_b^* + \langle \eta_b^* \rangle, \tag{13.4}$$

where $\tilde{\eta}_b^*$ is the basin tide and $\langle \eta_b^* \rangle$ is the mean basin level, both measured with respect to mean sea level. Similarly, for the inlet velocity

$$u^* = \tilde{u}^* + \langle u^* \rangle. \tag{13.5}$$

In this equation, $\tilde{u}^*$ is the tidal velocity and $\langle u^* \rangle$ is the mean velocity resulting from river flow.

With the trial solution

$$\tilde{\eta}_b^* = \hat{\tilde{\eta}}_b^* \sin(t^* - \alpha), \tag{13.6}$$

it follows from Eq. (13.2) that

$$\tilde{u}^* = \hat{\tilde{u}}^* \cos(t^* - \alpha). \tag{13.7}$$

Referring to Appendix 13.A, the tidal velocity amplitude $\hat{\tilde{u}}^*$ and basin tidal amplitude $\hat{\tilde{\eta}}_b^*$ are

$$\hat{\tilde{u}}^* = \hat{\tilde{\eta}}_b^* = \sqrt{\frac{\sqrt{\left(1 - K_2^2\right)^4 + \frac{4k_{10}^2}{K_1^4}} - \left(1 - K_2^2\right)^2}{\frac{2k_{10}^2}{K_1^4}}}, \tag{13.8}$$

and the phase α given by

$$\alpha = \tan^{-1}\left(\frac{k_{10}\hat{\tilde{u}}^*}{K_1^2\left(1 - K_2^2\right)}\right). \tag{13.9}$$

The expressions for the mean basin level and mean inlet velocity are, respectively,

$$\langle \eta_b^* \rangle = -\frac{1}{K_1^2} k_{00} \hat{\tilde{u}}^{*2}, \tag{13.10}$$

and

$$\langle u^* \rangle = -Q^*. \tag{13.11}$$

The coefficients k_{10} and k_{00} depend on the ratio $\langle u^* \rangle / \hat{\hat{u}}^*$, which makes the solution implicit. Also, as shown in Appendix 13.A, the solution is limited to values of $\langle u^* \rangle / \hat{\hat{u}}^* \leq 1$. For zero river flow the expression for k_{10} reduces to that for the Öszoy–Mehta Solution, i.e., $k_{10} = 8/3\pi$ (compare Eqs. (6.25) and (13.8)) and $k_{00} = 0$.

Escoffier and Walton (1979) use a slightly different approach to solve Eqs. (13.2) and (13.3). The difference with the expanded Öszoy–Mehta Solution is the treatment of the nonlinear bottom friction term. However, after rearranging some of the expressions in Escoffier and Walton (1979), the resulting solution is similar to the expanded Öszoy–Mehta Solution. The solution for the amplitude of the basin tide $\hat{\hat{\eta}}_b^*$ and velocity $\hat{\hat{u}}^*$ is the same as Eq. (13.8), with k_{10} replaced by a coefficient e. The solution for the basin water level set-up $\langle \eta_b^* \rangle$ is the same as Eq. (13.10) with k_{00} replaced by a coefficient f. The coefficients e and f are functions of the ratio $\langle u^* \rangle / \hat{\hat{u}}^*$ that are numerically calculated and presented in Fig. 1 of Escoffier and Walton (1979). Just as for the expanded Öszoy–Mehta Solution, the solution requires iteration and is limited to values of $\langle u^* \rangle / \hat{\hat{u}}^* \leq 1$.

To demonstrate the effect of the river flow on the basin water level and the inlet velocity, the expanded Öszoy–Mehta Solution is applied to the representative inlet. Parameter values for this inlet are given in Table 6.2. The results of the computations for the basin water level set-up and the amplitudes of the basin tide and the tidal velocity are presented in Fig. 13.1. The basin water level set-up increases and the amplitudes of the basin tide and the tidal velocity decrease with increasing river discharge.

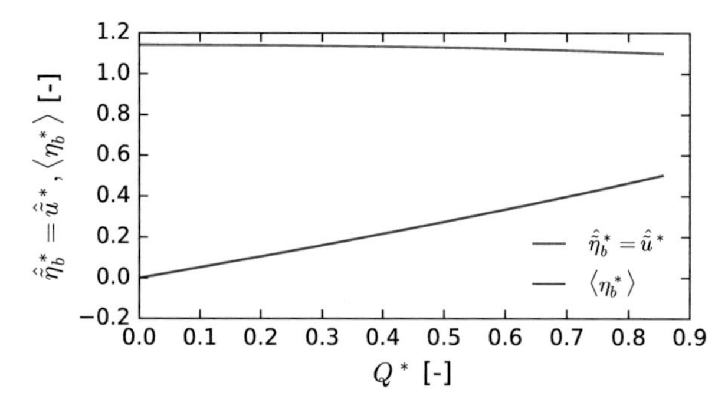

Figure 13.1 Basin water level set-up $\langle \eta_b^* \rangle$, amplitude of the basin tide $\hat{\hat{\eta}}_b^*$ and tidal velocity $\hat{\hat{u}}^*$ as a function of the river discharge Q^*. For parameter values, reference is made to Table 6.2.

13.3 Effect of River Flow on Cross-Sectional Stability of Selected Inlets

13.3.1 Thuan An Inlet: A Permanently Open Inlet

Thuan An Inlet is located on the central coast of Vietnam in a tropical monsoon region (Lam, 2009; Tung, 2011). The inlet opened in 1897 as a result of a breach in the barrier island. It was briefly closed in 1903–1904 but has been open ever since. Ocean tides are semi-diurnal with a tidal range of 0.41 m. The inlet connects the Tam Giang Lagoon to the Gulf of Tonkin. The part of the lagoon that is served by Thuan An Inlet has an estimated surface area of 104 km^2. Reported values of the tidal prism range from 28×10^6 m^3 to 47×10^6 m^3. The monsoon regime exerts its influence on the tidal inlet through the river flow. Most of this is concentrated in the period from September to December. Little is known about its magnitude, other than that the river flow exiting through Thuan An Inlet during the extreme flood of 1999 was estimated at 12,000 m^3 s^{-1}. Observed velocities in the inlet are never higher than 0.5 m s^{-1} (it is not clear whether this included periods of river flow). The mean annual offshore significant wave height is 1 m.

Estimates of the longshore sand transport at Thuan An Inlet vary between 0.4×10^6 m^3 $year^{-1}$ and 1.7×10^6 m^3 $year^{-1}$. With the reported values of the tidal prism, this result in P/M values ranging from 16 to 117. Taking an average value of 65 would imply that the location stability for tide alone is fair (Tables 3.1 and 3.2).

Entrance cross-sectional areas of Thuan An Inlet range from 2,900 to 4,000 m^2. Observations show that during the rainy season the inlet becomes wider and deeper. For example, as a result of the severe flood of November 1999 the inlet widened to 350 m and reached a maximum depth of 12 m. During the subsequent dry season, the inlet became narrower and shallower, a result of longshore sediment transport entering the inlet.

The cross-sectional stability of Thuan An Inlet is determined by both tide and river flow. To determine their relative importance, the cross-sectional stability for tide alone and tide and river flow is evaluated using the Escoffier Diagram introduced in Section 8.2. Parameter values for the inlet are presented in Table 13.1. The closure curves in the diagram are calculated using the expanded Öszoy–Mehta Solution. Because with river flow it is not possible to define a tidal prism, the velocity amplitude is taken as the maximum ebb velocity instead of the velocity $\hat{u}$ defined by Eq. (5.6). Assumed is a trapezoidal cross-section with side slopes 1:3 and a ratio of surface width to bottom width of 130, resulting in a shape factor $\beta_2 = 0.17$. Evolving cross-sections remain geometrically similar. Accounting for the moderate to large longshore sand transport, the equilibrium velocity $\hat{u}_{eq}$ is estimated to be on the high side of the values presented in Section 5.2.4 and taken equal to 1 m s^{-1}.

Table 13.1 *Parameter values for Thuan An Inlet used to calculate the closure curve.*

Parameter	Symbol	Dimension	Value
Inlet length	L	m	4,000
Friction factor	F	–	3×10^{-3}
Entrance/exit loss coefficient	m	–	1
Shape factor	β_2	–	0.17
Basin surface area	A_b	m^2	1.04×10^{8}
Tidal amplitude	$\hat{\eta}_0$	m	0.2
Tidal frequency	σ	rad s^{-1}	1.4×10^{-4}
Equilibrium velocity	$\hat{u}_{eq}$	m s^{-1}	1

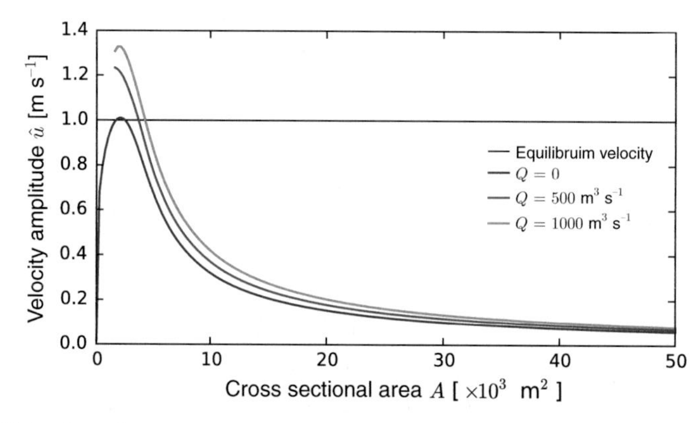

Figure 13.2 Escoffier Diagram for Thuan An Inlet with $Q = 0$, $Q = 500$ m^3 s^{-1} and $Q = 1,000$ m^3 s^{-1}. For parameter values, see Table 13.1.

The Escoffier Diagram with river flows of $Q = 0$, $Q = 500$ m^3 s^{-1} and $Q = 1,000$ m^3 s^{-1} is presented in Fig. 13.2. From this, it is concluded that for tide alone the stable equilibrium cross-sectional area is approximately 2,400 m^2. The unstable equilibrium cross-sectional area is only slightly less, implying that the degree of cross-sectional stability is marginal and chances are that in the absence of river flow the inlet closes. The stable equilibriums are considerably larger when including river discharge and are 3,600 m^2 for a river discharge of 500 m^3 s^{-1} and 4,200 m^2 for a river discharge of 1,000 m^3 s^{-1}.

During periods of river discharge, the inlet cross-sectional area increases beyond that for tide alone, thus increasing the entrance stability. When river flow diminishes, the inlet cross-sectional area decreases and tends towards the stable cross-sectional area for tide alone. In spite of the marginal stability for tide alone,

Thuan An Inlet is open year round. This is attributed to the longshore sand transport not being large enough to close the inlet before the next flood arrives.

13.3.2 Wilson Inlet: A Seasonally Open Inlet

Wilson Inlet is a small inlet on the southwest coast of Australia. Tides are diurnal with a spring tidal range of 0.8 m. The surface area of the back-barrier lagoon is approximately 25 km^2. The climate is characterized by a clearly defined wet and dry season, resulting in a seasonally varying river flow. The wet season coincides with the Australian winter (July–October). The mean annual river discharge is 207×10^6 m^3, of which 80 percent enters during the wet season. At the beginning of the wet season, the inlet is artificially opened by dredging a small channel through the entrance bar. Subsequently, it is kept open by river flow. Following the wet season, the inlet closes for a period of six to seven months due to the formation of a sand bar across the entrance. The coast has little exposure to wind waves limiting the longshore sand transport to an estimated 10,000 m^3 $year^{-1}$.

To study the closure of the inlet, a field study was carried out (Ranasinghe and Pattiaratchi, 1999). Field observations at the end of the wet season included bathymetric surveys and wave and current measurements. During the observational period the inlet was open with river flow being insignificant. Wave measurements showed waves to be dominantly swell that approached the coast perpendicularly. Longshore currents were observed to be weak. Current measurements showed maximum flood velocities of 1 m s^{-1} and maximum ebb velocities of 0.8 m s^{-1}. The predominant swell conditions, the weak and inconsistent longshore currents and the absence of spit formation suggest that for Wilson Inlet, onshore transport of sand is the main transport process responsible for inlet closure.

This was further investigated using a morphodynamics model (Ranasinghe et al., 1999). The overall structure of the morphodynamics model is similar to that presented in Chapter 11, with separate modules for flow, waves, sand transport and bathymetric changes. Applying the model to Wilson Inlet, it was found that, in the absence of river flow, longshore sand transport alone would not close the inlet, whereas swell-generated onshore sand transport alone would. This confirms the conclusion from the field observations that onshore sand transport, rather than longshore sand transport, is responsible for the seasonal closure of Wilson Inlet.

13.3.3 Lake Conjola Inlet: An Intermittently Open Inlet

Lake Conjola Inlet is a small inlet on the southeastern coast of Australia. Tides are semi-diurnal and at the upper end of micro-tidal. River flow is intermittent. The coast is characterized by high energy waves. Longshore sand transport is variable both in magnitude and direction. Onshore sand transport may be at least as important as longshore sand transport in carrying sand towards the inlet (McLean and Hinwood, 2000).

In the absence of river flow, Lake Conjola Inlet has an inadequate tidal prism to remain open and continually moves to a closed state. Two means of entrance restrictions have been observed: 1) gradual closure through inward tidal transport of sand and 2) a sudden restriction associated with coastal storms, resulting in an entrance bar (Hinwood and McLean, 2001). Summarizing, entrance conditions are sensitive to coastal storms and fluvial events with tidal flows providing the energy producing gradual entrance change following these large perturbations.

The dependence of the entrance condition on fluvial and storm events, rather than on a trend towards a regime state, is illustrated using the two diagrams in Fig. 13.3. The diagrams show non-dimensional values of basin tidal amplitude $\hat{\hat{\eta}}_b*$ (Fig. 13.3a) and basin water level set-up $\langle \hat{\eta}_b^* \rangle$ (Fig. 13.3b) as a function of non-dimensional river discharge Q^* and repletion coefficient K_1. Neglecting inertia,

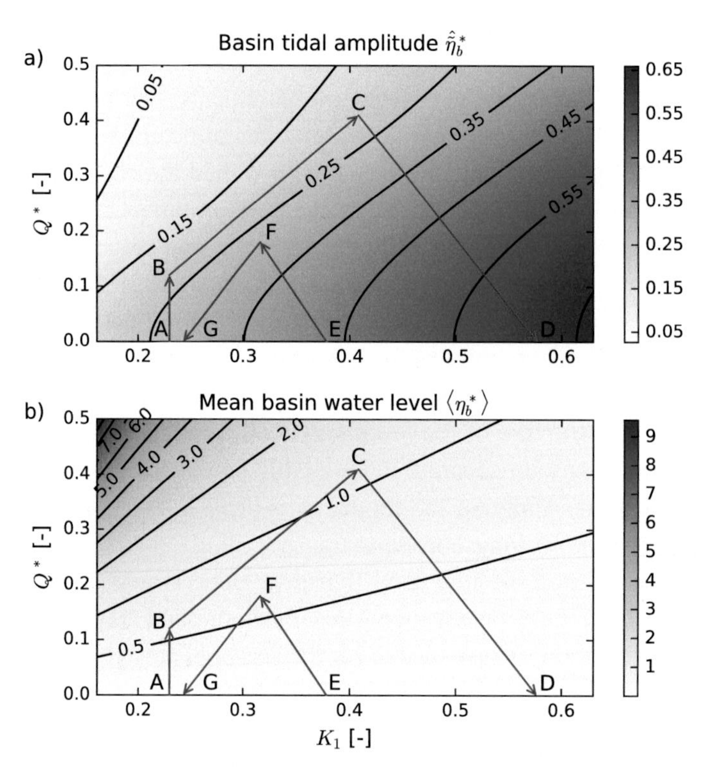

Figure 13.3 a) Basin tidal amplitude and b) mean basin water level as a function of Q^* and K_1 with trajectories (in red) for the response of Lake Conjola Inlet to a fluvial event (ABCD) and a storm event (EFG) (adapted from Hinwood and McLean, 2001).

the diagrams are constructed using Eqs. (13.1) and (13.2) with $K_2 = 0$. The equations are solved numerically (van de Kreeke, 1967). By calculating Q^* and K_1 for typical fluvial and storm events and including them in the diagram, the effect of these events on basin tidal amplitude and basin water level set-up may be seen as trajectories in the diagrams.

In case of a fluvial event, Q^* initially increases but the entrance cross-sectional area has insufficient time to substantially increase, hence the value of K_1 does not change. The trajectory is shown as AB, which shows an increased mean basin level and some attenuation of the tidal amplitude. When river flow increases, it continues to increase Q^* and enlarges the entrance, thereby increasing K_1. The corresponding trajectory is BC. At C the mean basin water level is raised and the basin tidal amplitude is slightly reduced. The attenuation of the tidal amplitude is a result of the increased river flow which is only partially offset by a decrease in the resistance parameter. When the river flow diminishes, Q^* decreases and, with the entrance cross-sectional area still increasing, the value K_1 increases. This results in trajectory CD. While at D, the mean basin water level approaches the mean sea level and the basin tidal amplitude has increased compared to the value before the river flow event.

In the case of a storm event, the steep waves carry sand into the inlet, decreasing the value of K_1. Storms are usually accompanied by some rainfall, resulting in an increase in Q^*. A typical trajectory for such an event is EF. Because of the increased river flow and the increase in the resistance parameter, at F the basin tidal amplitude has decreased. After the river flow decreases, the entrance constriction remains. Following trajectory FG, as a result of the decreasing river flow, the tidal amplitude increases and the mean basin water level decreases. When at G, the basin tidal amplitude is smaller than at the beginning of the event and the mean tide level is the same, approaching mean sea level.

13.4 A Morphodynamic Model for the Long-Term Evolution of an Inlet

To improve the understanding of the long-term evolution of an inlet in the presence of tide and river flow, Hinwood et al. (2012) and Hinwood and McLean (2015a) developed a process-based exploratory morphodynamics model. The model consists of a hydrodynamics, a sediment transport and a bathymetry module. Schematization of the tidal inlet is shown in Figs. 6.1 and 6.2. The hydrodynamics is described by Eq. (7.1) and includes the variation of depth with tidal stage. With the water level in the inlet approximated by the basin tide, Eq. (7.1) is written as

$$\frac{L}{g}\frac{du}{dt} + \left(\frac{m}{2g} + \frac{FL}{g(h+\eta_b)}\right)u|u| = \eta_0 - \eta_b. \tag{13.12}$$

Assumed is a rectangular inlet cross-section with a tidally averaged depth h; L is inlet length, g is gravity acceleration, u is cross-sectionally averaged velocity, positive in the (tidal) flood direction, t is time, m is entrance/exit loss coefficient, F is a friction factor, η_0 is ocean tide and η_b is basin tide. Continuity is described by Eq. (7.2), with a term added to account for the river flow and, as in Eq. (13.12), the water level in the inlet approximated by the basin tide, resulting in

$$b(h + \eta_b)u = A_b \frac{d\eta_b}{dt} - Q, \tag{13.13}$$

where b is inlet width, A_b is basin surface area and Q is river discharge.

The ocean tide is a simple harmonic

$$\eta_0 = \hat{\eta}_0 \sin(\sigma t), \tag{13.14}$$

where $\hat{\eta}_0$ is ocean tidal amplitude and σ is the tidal frequency. In the sediment transport module, sediment transport is in the form of suspended load with C_0 and C_b the sediment concentrations in the ocean and the basin, respectively. The concentration in the inlet follows from

$$\begin{aligned} C &= k\left[\left(\frac{u}{u_c}\right)^2 - 1\right], \quad &\text{for } u \geq u_c, \\ C &= 0, &\text{for } u < u_c. \end{aligned} \tag{13.15}$$

In this equation, C is the sediment concentration in the inlet, u_c is a threshold velocity for both pickup and deposition and k is an empirical constant. During flood, the sand transport into the inlet is $b(h + \eta_0)|u|C_0$ and the transport out of the inlet is $b(h + \eta_0)|u|C$. During ebb, the sand transport into the inlet is $b(h + \eta_b)|u|C_b$ and the transport out of the inlet is $b(h + \eta_b)|u|C$.

The bathymetry module consists of the conservation of sediment equation,

$$L\frac{dh}{dt} = (h + \eta_0)|u|(C - C_0), \qquad \text{for flood}, \tag{13.16}$$

and

$$L\frac{dh}{dt} = -(h + \eta_0)|u|(C - C_b), \qquad \text{for ebb}. \tag{13.17}$$

For flood, there is erosion for $C_0 < C$ and deposition for $C_0 > C$. For ebb, there is erosion for $C_b < C$ and deposition for $C_b > C$.

Except for the wave module, computations follow the feedback loop described in Section 11.2. Starting with an initial value for the tidally averaged depth, h, Eqs. (13.12) and (13.13) with the boundary condition Eq. (13.14) are solved numerically for $u(t)$ and $\eta_b(t)$. Concentrations are then calculated from Eq. (13.15) and used in Eqs. (13.16) and (13.17) to calculate the change in depth. Some parameter values, notably C_0 and C_b, are based on observations in intermittently closed inlets on Australia's southeast coast. Others, like k, F and u_c, are taken from the literature.

The relatively simple formulation of the model makes it possible to carry out many calculations in a short period of time. As an example, for a given ocean tide and river flow, calculations were made for a large number of initial inlet depths. Runs were made for a sufficiently long duration to show that after some time the depth reaches an equilibrium value. Results are plotted in a $\frac{\pi}{2}Q^* - h^*$ diagram (Fig. 13.4), with the non-dimensional river flow, Q^*, given by Eq. (13.3) and the non-dimensional water depth $h^* = h/\hat{\eta}_0$. The expression $\frac{\pi}{2}Q^*$ represents the non-dimensional river flow used by Hinwood and McLean (2015a). The left panel in Fig. 13.4 shows the starting positions, which are evenly distributed over the $\frac{\pi}{2}Q^* - h^*$ plane. The right panel shows the final position of each run. Final positions, representing the equilibrium depths, are clustered in two narrow bands, referred to as tidal flow and river flow attractors. The equilibrium depths depend on the value of river flow but are independent of the initial depth. An initially shallow inlet will evolve towards the attractor by increasing its depth and an initially deep inlet will evolve towards the attractor by decreasing its depth. Even though the model shows the attractors to cluster in two bands, a physical explanation is still lacking.

For zero river flow the model described in this section leads to similar results as the Escoffier Stability Model, and inlets approach an equilibrium depth that is independent of the initial depth. Because of the different sand transport formulations, the values of the equilibrium depths are expected to be different for the two models (Hinwood and McLean, 2015b).

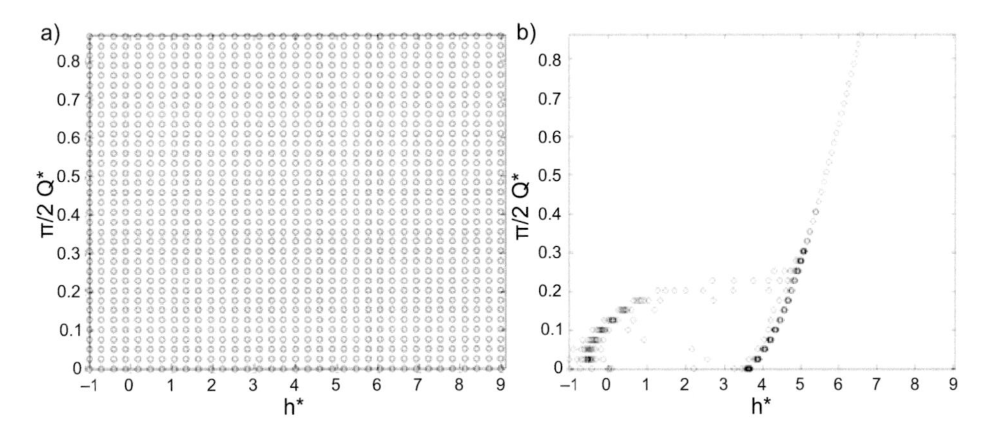

Figure 13.4 Evolution of inlet depths. a) The starting conditions for the model runs and b) the final depth positions showing clustering in two narrow bands (reprinted from Hinwood and McLean, 2015b, copyright 2016, with permission from Elsevier).

13.A Öszoy–Mehta Solution Including River Flow

The governing equations are Eqs. (13.1) and (13.2) in the main text. Assumed is a sinusoidal ocean tide

$$\eta_0^* = \sin(t^*), \tag{13.A.1}$$

with the non-dimensional variables defined by Eq. (6.5).

The solution for the basin water level, η_b^*, is written as the sum of the basin tide $\tilde{\eta}_b^*$ and a mean basin water level $\langle \eta_b^* \rangle$,

$$\eta_b^* = \tilde{\eta}_b^* + \langle \eta_b^* \rangle. \tag{13.A.2}$$

Similarly, the inlet velocity is written as the sum of a tidal velocity $\tilde{u}^*$ and a mean velocity $\langle u^* \rangle$,

$$u^* = \tilde{u}^* + \langle u^* \rangle. \tag{13.A.3}$$

Substituting for η_b^* and u^* from, respectively, Eqs. (13.A.2) and (13.A.3) in Eq. (13.2) and collecting time-independent and time-dependent terms results in, respectively,

$$\langle u^* \rangle = -Q^*, \tag{13.A.4}$$

and

$$\tilde{u}^* = \frac{d\tilde{\eta}_b^*}{dt^*}. \tag{13.A.5}$$

With trial solutions for the basin tide and tidal velocity of, respectively,

$$\tilde{\eta}_b^* = \hat{\tilde{\eta}}_b^* \sin(t^* - \alpha), \tag{13.A.6}$$

and

$$\tilde{u}^* = \hat{\tilde{u}}^* \cos(t^* - \beta), \tag{13.A.7}$$

it follows from Eq. (13.A.5) that

$$\hat{\tilde{\eta}}_b^* = \hat{\tilde{u}}^* \qquad \text{and} \qquad \beta = \alpha. \tag{13.A.8}$$

To arrive at a solution of Eqs. (13.1) and (13.2), the product $u^*|u^*|$ is expanded in a Fourier series. Following Dronkers (1964, 1968), the Fourier expansion is carried out by substituting the expression for u^*, Eq. (13.A.3) with $\tilde{u}^*$ given by Eq. (13.A.7), in $u^*|u^*|$. Retaining only the constant and first harmonic in the series expansion results in

$$u^*|u^*| \cong k_{00}\hat{\tilde{u}}^{*2} + k_{10}\hat{\tilde{u}}^{*2} \cos(t^* - \beta), \tag{13.A.9}$$

with

$$k_{00} = \frac{1}{4}(2 + 2\gamma)\left(2 - \frac{4\gamma}{\pi}\right) + \frac{2}{3\pi}\sin(2\gamma), \tag{13.A.10}$$

$$k_{10} = \frac{3}{\pi}\sin(\gamma) + \frac{1}{3\pi}\sin(3\gamma) + \left(2 - \frac{4\gamma}{\pi}\right)\cos(\gamma), \tag{13.A.11}$$

and

$$\cos(\gamma) = \frac{\langle u^* \rangle}{\hat{\hat{u}}^*}. \tag{13.A.12}$$

For zero river flow with $\gamma = \pi/2$, $k_{00} = 0$ and $k_{10} = 8/(3\pi)$. With $\langle u^* \rangle$ negative, γ is in the second quadrant. Eq. (13.A.12) requires that $\langle u^* \rangle < \hat{\hat{u}}^*$. It follows from Eqs. (13.A.10), (13.A.11) and (13.A.12) that both k_{00} and k_{10} are functions of the unknown $\hat{\hat{u}}^*$.

Using the Fourier expansion of $u^*|u^*|$, and substituting for η_0^*, η_b^* and u^*, from, respectively, Eqs. (13.A.1), (13.A.6) and (13.A.7) in Eq. (13.1) and using Eqs. (13.A.6), (13.A.7) and (13.A.8), it follows that:

$$\left(1 - K_2^2\right)\hat{\hat{u}}^* \sin(t^* - \alpha) + \frac{k_{00} + k_{10}\cos(t^* - \alpha)}{K_1^2}\hat{\hat{u}}^{*2} = \sin(t^*) - \langle \eta_b^* \rangle. \tag{13.A.13}$$

Collecting time-dependent terms results in an equation for the tidal velocity amplitude:

$$\left(1 - K_2^2\right)\hat{\hat{u}}^* \sin(t^* - \alpha) + \frac{k_{10}}{K_1^2}\hat{\hat{u}}^{*2}\cos(t^* - \alpha) = \sin(t^*). \tag{13.A.14}$$

Collecting time-independent terms results in the equation for the mean basin level:

$$\langle \eta_b^* \rangle = -\frac{k_{00}}{K_1^2}\hat{\hat{u}}^{*2}. \tag{13.A.15}$$

For zero river flow, with $k_{00} = 0$, the basin water level set-up is zero.

Except for the factor $8/3\pi$, which is replaced by k_{10}, Eq. (13.A.14) is the same as Eq. (6.20). Using the same method used to solve Eq. (6.20), the solution to Eq. (13.A.14) is

$$\hat{\hat{u}}^* = \sqrt{\frac{\left[\left(1 - K_2^2\right)^4 + \frac{4k_{10}^2}{K_1^4}\right]^{1/2} - \left(1 - K_2^2\right)^2}{\frac{2k_{10}^2}{K_1^4}}}, \tag{13.A.16}$$

and

$$\alpha = \tan^{-1}\left(\frac{k_{10}\hat{\hat{u}}^*}{K_1^2\left(1 - K_2^2\right)}\right). \tag{13.A.17}$$

With k_{10} a function of the velocity amplitude, Eq. (13.A.16) is implicit in $\hat{\hat{u}}^*$ and has to be solved iteratively. Once $\hat{\hat{u}}^*$ is known, values of $\langle \eta_b^* \rangle$, $\hat{\hat{\eta}}_b^*$ and α follow from Eqs. (13.A.15), (13.A.8) and (13.A.17), respectively. For zero river flow, with $k_{10} = 8/3\pi$, Eqs. (13.A.16) and (13.A.17) are the same as Eqs. (6.25) and (6.26), respectively.

14

Engineering of Tidal Inlets

14.1 Introduction

Back-barrier lagoons are home to recreational marinas and fishing ports. With a few exceptions, most vessels using these facilities are relatively small, with lengths in the 5–30 m range and a maximum draft of 5 m. To access the lagoon, vessels need to navigate the ebb delta channel and the inlet. This requires that both channel and inlet are relatively stable, have sufficient depth and an alignment relative to the wave direction that allows safe access and passage. Not many natural inlets satisfy these requirements and measures are needed to remedy the shortcomings. A distinction is made between soft and hard measures. Soft measures include the opening of a new inlet, inlet relocation, dredging and artificial sand bypassing. Hard measures are jetty construction and weir-jetty systems. In addition to providing boating access, inlets play a role in maintaining the water quality of the back-barrier lagoon; they serve as conduits for the exchange of lagoon and ocean water.

14.2 Artificial Opening of a New Inlet

The objectives of the artificial opening of a new inlet are to provide passage for vessels to the back-barrier lagoon and/or to improve water quality. With regards to the passage of vessels, design requirements include sufficient channel depth, width, alignment and stability. Improving water quality requires sufficient exchange, implying a large enough tidal prism. Examples of inlets that were artificially opened, but with different objectives, are Bakers Haulover Inlet (FL), Faro-Olhão Inlet (Portugal) and Packery Channel (TX). Bakers Haulover Inlet (Fig. 14.1a) was opened in 1925 for the prime purpose of improving water quality in the northern part of Biscayne Bay (Dombrowsky and Mehta, 1993). Faro-Olhão Inlet (Fig. 14.1b) was opened in 1929 to improve navigational access to the city of Faro

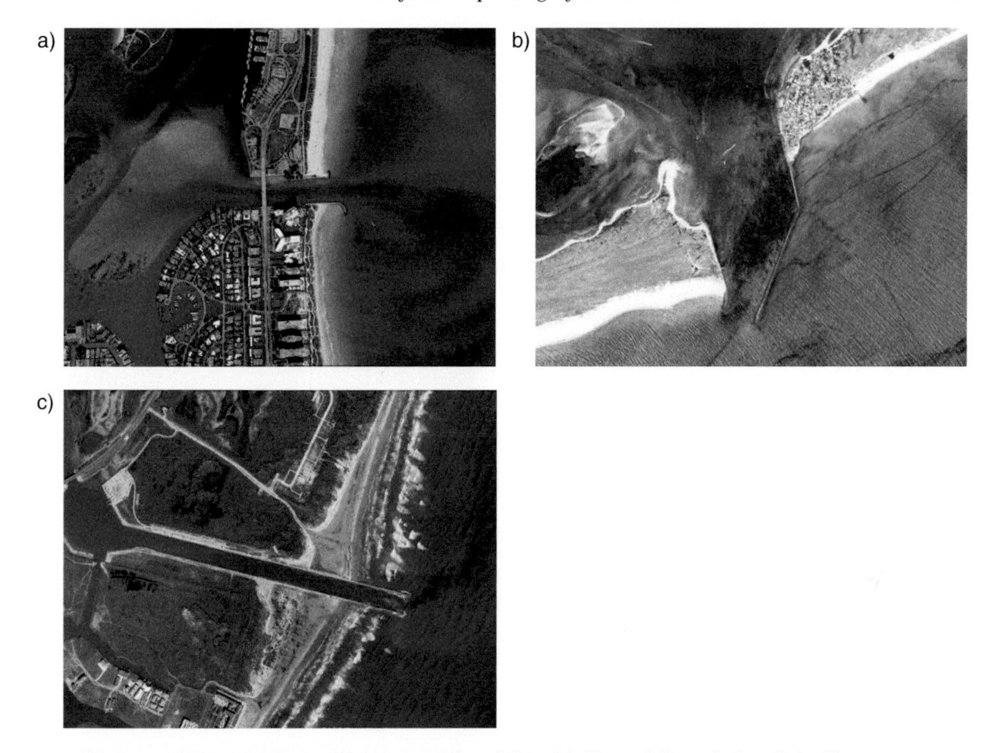

Figure 14.1 a) Bakers Haulover Inlet (FL), b) Faro-Olhão Inlet (Ría Formosa, Portugal) and c) Packery Channel (TX) (Source: Google Earth).

(Pacheco et al., 2011). Packery Channel (Fig. 14.1c) was opened in 2006 with the objectives to facilitate recreational fishing and boating and to improve the exchange between the Gulf of Mexico and Corpus Christi Bay (Williams et al., 2007).

In the following, some of the elements that go into the design of a new inlet are discussed. A first step is to determine the equilibrium entrance cross-sectional area and its stability. For this the Escoffier Diagram, described in some detail in Section 8.2.2, is used. Assuming the back-barrier lagoon is small and deep enough to allow for a uniformly fluctuating basin water level, the closure curve in the diagram can be calculated using the Öszoy–Mehta or the Keulegan Solution (Sections 6.4 and 6.5). For back-barrier lagoons that are large and shallow, more advanced models should be used. This implies an increase in computer time. To limit computer time and cost, a solution is to run the more advanced models in combination with a simpler, less computationally extensive model. The simple model could then be used to narrow down the parameter range of interest and reduce the computing time of the advanced model.

Calculating the equilibrium velocity curve in the Escoffier Diagram requires that the A–P relationship for the location of the new inlet is known. If not available, the next best option is to look for relationships for similar coasts, i.e., coasts with the same offshore wave characteristics (wave height, period and direction), same tide characteristics (semi-diurnal, diurnal or mixed tide) and the same grain characteristics (density and size). Depending on the selected $A–P$ relationship, the equilibrium velocity is given by Eq. (5.12) or Eq. (5.13). The equilibrium velocity is usually close to 1 m s^{-1}, with values slightly larger for coasts with a large longshore sand transport and somewhat smaller for coasts with a small longshore sand transport.

With the known stable cross-sectional area determined from the Escoffier Diagram, the tidal prism follows from the cross-sectional area – tidal prism relationship. With the known tidal prism, P, and assuming the gross longshore sand transport M is known, the P/M ratio is calculated. Using Tables 3.1 and 3.2, this gives the degree of location stability and the bypassing mode for the new inlet. To assure sufficient stability of the inlet and the channels on the ebb delta, a P/M ratio larger than 80 is recommended.

After opening a new inlet, ebb and flood deltas develop. The deltas continue to grow by capturing part of the longshore sand transport, until they reach equilibrium. An example is Ocean City Inlet (MD), for which the development of the ebb delta was observed and calculated using the empirical model discussed in Section 12.3. Without the new inlet, the sand stored in the deltas would have been available to the downdrift beach. Given the tidal prism, an estimate of the ebb delta volume, and the potential sand loss to the downdrift beach, follows from Eq. (5.14). In particular, for inlets with a large tidal prism the sand volume of the deltas is substantial, resulting in potentially large erosion of the downdrift beaches. To partly compensate for the sand loss to the downdrift beach, it is recommended that sand from opening the inlet is placed on the downdrift beach or at the future location of the ebb delta.

An important step in determining the effect of a new inlet on the downdrift shore is development of a sand management plan. The sand management plan includes the sediment budget discussed in Section 3.2. For a given control volume, it delineates the sources and sinks of sediment, the sediment transport, sediment transport pathways and volume changes. The sediment budget should be constructed for both the adaptation/transition period and the period after the morphology of the tidal inlet has reached equilibrium. A helpful tool in constructing the sediment budget, especially for the adaptation/transition period, is the empirical models described in Chapter 12. For most inlets, the major source of sand is the longshore sand transport. Major sinks are the back-barrier lagoon and the offshore. The sediment budget should result in an estimate of the volume of sand that bypasses the inlet at any given period and reaches the downdrift coast. Given this volume the resulting shoreline changes can be calculated with the line models described in Kamphuis

(2006) and Bakker (2013). Examples of inlet sand management plans for eight inlets along the southeast coast of Florida are presented in Dombrowsky and Mehta (1993).

Results of preliminary studies should indicate whether inlet cross-sectional area meets navigational demands, a sufficient volume of sand is bypassed to the downdrift beach and cross-sectional and location stability are large enough to limit maintenance dredging. If any of these requirements is not met, measures described in Sections 14.3–14.7 can be contemplated.

14.3 Relocation of an Existing Inlet

Inlets with a relatively small P/M ratio have a tendency to migrate and form a sand spit across the entrance. Spit formation increases the inlet length and leads to a lowering of the inlet velocity and shoaling. To improve navigability and/or water exchange at these inlets, one measure is to relocate the inlet. Examples are Captain Sam's Inlet and Mason Inlet, discussed in Sections 4.4 and 4.5, respectively. Another well-documented example is Ancão Inlet (Pacheco et al., 2007; Vila-Concejo et al., 2003). Ancão Inlet is part of the Ría Formosa barrier island system in Portugal. The inlet has a history of eastward migration and breaching. In late 1996, the inlet was in the last stage of its migration cycle, highly sinuous and infilling. To assure navigability and to provide sufficient water exchange, in the middle of 1997, a new inlet was opened 3.5 km to the west of the closed inlet. After a relatively rapid adjustment, the relocated inlet started to resemble a natural inlet with well-developed ebb and flood deltas. After moving a short distance to the west it resumed its migration towards the east. The inlet in 2004 (Fig. 14.2a) still showed the classic features of a natural inlet, but in 2011 (Fig. 14.2b) had deteriorated to the extent that artificial breaching again was considered. The effect of the inlet relocation was felt up to 4 km updrift of the opening position, showing an

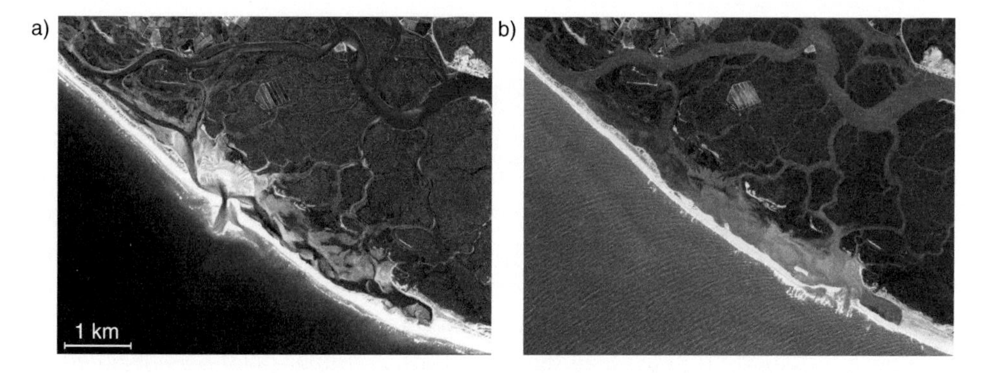

Figure 14.2 Ancão Inlet in a) 2004 and b) 2011 (Source: Google Earth).

increase in beach volume (Ferreira, 2011). Erosion after relocation was observed downdrift of the relocation position (Matias et al., 2009). Similar the dredging of shoals and shallows (Section 14.4) and sand bypassing (Section 14.5), relocation of a migrating inlet is a repeat process. Partly based on the studies carried out at Ancão Inlet, recommendations for inlet relocation are presented in Vila-Concejo et al. (2004).

14.4 Dredging

Dredging at inlets includes maintenance dredging and sand bypassing. Maintenance dredging refers to the removal of shoals and shallows from inlet and navigation channels. Sand bypassing involves dredging sand from a sand trap or impoundment basin and moving it to the downdrift coast. Sand traps or impoundment basins are pre-dredged depressions where sand is collected by natural processes. They are usually located in protected waters that allow dredges to operate in a quiet wave environment.

Dredges used at inlets are hydraulic suction dredges and hopper dredges (Vlasblom, 2003). In case of a hydraulic suction dredge, sand is transported through a floating or submerged pipeline onto the downdrift beach or it is offloaded on a split hull barge and dumped on the foreshore. Hopper dredges carry the sand to the downdrift beaches where they dump it on the foreshore or pump it onto the beach by a floating pipe line or by rainbowing. For details of these methods reference is made to the US Army Corps of Engineers' *Shore Protection Manual* (1984).

An example of an inlet that is maintained by dredging is Wiggins Pass (FL). Using a hydraulic suction dredge, the sand is directly placed on the downdrift beach via a pipeline (Dabees et al., 2011). Examples of inlets with sand traps or impoundment basins from which the sand is transported to the downdrift beaches are Sebastian Inlet (FL), Jupiter Inlet (FL), Masonboro Inlet (NC) and Channel Islands Harbor (CA).

14.5 Sand Bypassing Plants

Bypassing plants are designed to transfer sand accumulated at the updrift jetty directly to the downdrift beaches, avoiding the need for an impoundment basin and dredging. The oldest sand bypassing plant is at South Lake Worth Inlet (FL) and was constructed in 1937. The plant uses slurry pumps that operate from a platform on the updrift jetty. A submerged discharge line carries the sand–water mixture to the downdrift beach. After upgrading the pump in 1948, the design capacity was estimated at 70,000 m^3 $year^{-1}$. In the late fifties, a similar plant was installed at Lake Worth Inlet, located approximately 25 km to the north (Fig. 14.3). The plant was designed to bypass 230,000 m^3 $year^{-1}$. The sand–water mixture is carried

Figure 14.3 Sand bypassing plant at Lake Worth Inlet (FL) (Source: Google Earth).

through a submerged discharge line to the downdrift beach. A detailed description of the plant is given in (Zurmuhlen, 1957). A problem with fixed bypassing plants, as constructed at South Lake Worth Inlet and Lake Worth Inlet, is that sand does not always accumulate at the same location relative to the jetty. This, and the limited length of the dredge arm, means that the intake cannot always reach the accumulated sand. This makes the plants inefficient.

To deal with the natural variability of the location of the accumulated sand fillet, at Nerang River Entrance (Queensland, Australia) a 500 m long pier was constructed perpendicular to the beach close to the updrift jetty (Fig. 14.4). Along the outer 300 m, ten jet pumps were installed. As opposed to conventional slurry pumps, jet pumps do not have moving parts. At Nerang River Entrance the slurry from the jet pumps is discharged in a buffering hopper and from there is discharged to the downdrift beach. Operation started in 1986. The design bypass rate is 600,000 m^3 $year^{-1}$. A detailed description of the plant is given in US Waterway Experiment Station (1989).

In 1990, a moving bypassing system was implemented at Indian River Inlet (DE). The system uses a single jet pump. To account for the natural variability of the location of the accumulated sand, the jet pump must cover a relatively large area. This is accomplished by employing the jet pump from a crawler crane moving along the beach in the swash zone. Discharge of the jet pump is through a line to a booster pump and from there a discharge line crosses the inlet via an existing bridge. The discharge line extends up to a maximum of 500 m on the beach on the

Figure 14.4 Sand bypassing plant at Nerang River Entrance (Queensland, Australia) (Source: Google Earth).

downdrift beach. The system was designed to bypass 90,000 m^3 $year^{-1}$. For details of the operation and performance, reference is made to Clausner et al. (1991).

14.6 Jetties; Jetty Length and Orientation

Jetties are implemented to prevent the longshore sand transport from entering the navigation channel. In addition, they confine the current, thereby reducing maintenance dredging (Kraus, 2005). Inlets have one or two jetties. The more common configuration is that of two parallel jetties. Examples are presented in Figs. 14.1a and 14.1c. Where sand traps and wave damping beaches are located within the confines of the jetties, the configuration is more as shown in Fig. 14.1b.

To prevent sand from entering the navigation channel, jetties should be sand tight and preferably extend beyond the seaward limit of the longshore sand transport. When shorter, some of the longshore sand transport will enter the navigation channel. This sand has to be removed by dredging, which in some cases might be cheaper than extending the jetty. For the seaward limit of the longshore sand transport, reference is made to Komar (1998) and Kamphuis (2006).

After the sand fillet at the updrift side of a jetty has sufficiently developed, some sand travels seaward along the jetty and at the jetty tip enters the navigation channel. To slow this process, sometimes spur jetties are added (Fig. 14.5). The spur jetty is typically perpendicular to the jetty but could also be at an acute

Figure 14.5 Jetty with spur jetty at Fort Pierce Inlet(FL) (Source: Google Earth).

angle. The basic function of a spur jetty is to divert the sand, keeping it away from the jetty and navigation channel. For spur jetty design considerations including location, elevation and length, reference is made to Seabergh and Kraus (2003).

Depending on length and orientation, jetties contribute to the safe entrance of vessels into the inlet. Preferably, jetties should extend beyond the breaker zone. The location of the breaker zone depends on wave height, wave period, bottom slope and tidal stage. As a simple rule, waves break when the wave height approaches the water depth. As a result, on any given day, the distance between the shore and the breaker zone differs. The relevant breaker zone is where waves break for wave conditions at which the design vessel should still be able to enter the inlet. Summarizing, jetties should preferably extend beyond the seaward limit of the longshore sand transport and beyond the relevant breaker zone. In addition to jetty length, jetty orientation is important. To maintain rudder control, orientation of the jetties should be such that, under rough conditions, vessels can enter with waves coming in at the aft quarter as opposed to having to deal with a following sea.

Jetties interrupt the natural sand transport pathways (Section 3.3), leading to adverse effects on the downdrift beaches. Therefore, a central element of inlet restoration and improvement is the bypassing of the sand accumulated at the updrift side of the jetties. Preferably, bypassing is accomplished by natural processes, whereby sand is transferred to the downdrift beaches by waves and tide. If this is not possible, artificial means are used to bypass the sand. Examples are dredging

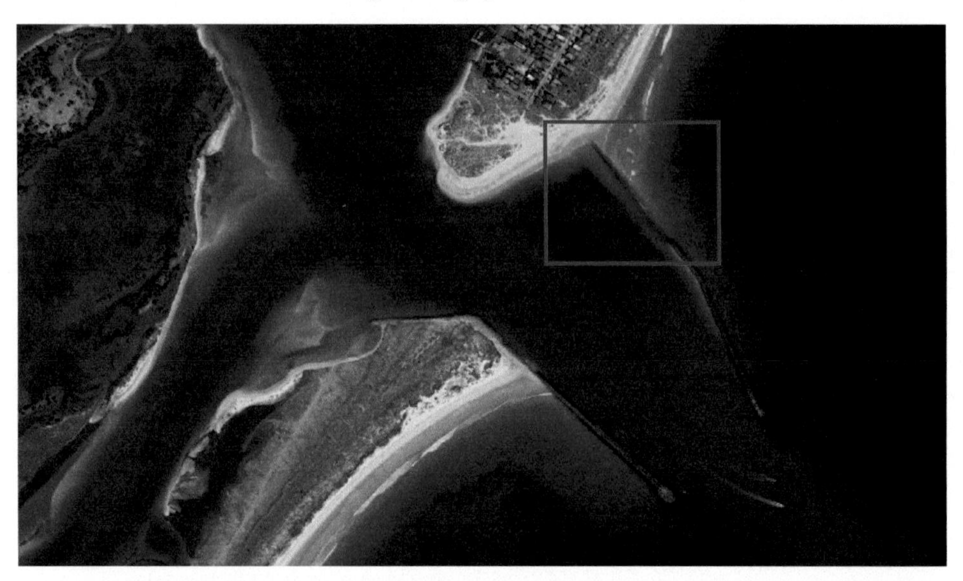

Figure 14.6 Weir-jetty system at Masonboro Inlet (NC) (Source: Google Earth).

(Section 14.4) and bypassing plants (Section 14.5). In the mid-sixties the weir-jetty system was introduced to bypass sand.

14.7 Weir-Jetty Systems

Jetties interrupt the flow of sand. A weir-jetty system allows the transport of sand across the jetty. The weir section is a depressed region in the jetty. The sand that is transported over the weir is collected in an impoundment basin or sand trap. The location of the impoundment basin or sand trap in the lee of the jetty allows small dredges to transfer the sand to the downdrift beaches. As an example, the weir-jetty system at Masonboro Inlet (NC) is shown in Fig. 14.6. An additional benefit of the weir-jetty system is that the seaward transport of sand along the outside of the jetty is reduced. As a result, the jetty may not need to extend as far as without a weir. Another benefit is that flood currents enter the inlet over the weir, bypassing the navigation channel. This leads to flood currents in the navigation channel being weaker than ebb currents, thereby reducing the volume of sand that enters the inlet.

The first weir-jetty systems were built in the mid-sixties at Hillsboro Inlet (FL) and Masonboro Inlet (NC). They were followed in the seventies and eighties by weir-jetty systems at Ponce de Leon Inlet (FL), Murrells Inlet (NC) and the mouth of the Colorado River (TX). A review of these projects, including their performance and recommendations for weir location, elevation and length, and location and size of the deposition basin, is presented in Seabergh and Kraus (2003).

References

Bakker, W.T. (1968). "A mathematical theory about sand waves and its application on the Dutch Wadden Isle of Vlieland". *Shore and Beach* 36(2): 4–14.
— (2013). *Coastal Dynamics*. World Scientific Publishing Co., p. 114.
Baldock, T.E., F. Weir, and M.G. Hughes (2008). "Morphodynamic evolution of a coastal lagoon entrance during swash overwash". *Geomorphology* 95(3–4): 398–411. DOI: 10.1016/j.geomorph.2007.07.001.
Batten, B.K., N.C. Kraus, and L. Lin (2007). "Long-term inlet stability of a multiple inlet system, Pass Cavallo, Texas". In: *Coastal Sediments '07*. American Society of Civil Engineers, pp. 1515–1528. DOI: 10.1061/40926(239)117.
Beets, D.J. and A.J.F. van der Spek (2000). "The Holocene evolution of the barrier and the back-barrier basins of Belgium and the Netherlands as a function of late Weichselian morphology, relative sea-level rise and sediment supply". *Netherlands Journal of Geosciences* 79(1): 3–16.
Biegel, E.J. and P. Hoekstra (1995). "Morphological response characteristics of the Zoutkamperlaag (the Netherlands) to a sudden reduction in basin area". *The International Association of Sedimentologists* 24: 85–99.
Booij, N., R.C. Ris, and L.H. Holthuijsen (1999). "A third-generation model for coastal regions: 1. Model description and validation". *Journal of Geophysical Research* 104(C4): 7649–7666. DOI: 10.1029/98JC02622.
Boon, J.D. and R.J. Byrne (1981). "On basin hypsometry and the morphodynamic response of coastal inlet systems". *Marine Geology* 40(1–2): 27–48. DOI: 10.1016/0025-3227(81)90041-4.
Boyd, J.P. (2001). *Chebyshev and Fourier Spectral Methods*. 2nd edn. New York: Dover Publications, 665 pp.
Brouwer, R.L. (2006). "Equilibrium and stability of a double inlet system". MA thesis. Delft University of Technology.
Brouwer, R.L. (2013). "Cross-sectional stability of double inlet systems". PhD thesis. Delft University of Technology.
Brouwer, R.L., H.M. Schuttelaars, and P.C. Roos (2013). "Modelling the influence of spatially varying hydrodynamics on the cross-sectional stability of double inlet systems". *Ocean Dynamics* 63(11): 1263–1278. DOI: 10.1007/s10236-013-0657-6.
Brouwer, R.L., H.M. Schuttelaars, J. van de Kreeke, and T.J. Zitman (2008). "Effects of amplitude differences on equilibrium and stability of a two-inlet bay system". In: *Proceedings of the 5th IAHR Symposium on River, Coastal and Estuarine Morphodynamics (RCEM) 2007*. Ed. by C. Dohmen-Janssen and S. Hulscher. Vol. 1. Leiden: Taylor & Francis/Balkema, pp. 33–39.

Brouwer, R.L., J. van de Kreeke, and H.M. Schuttelaars (2012). “Entrance/exit losses and cross-sectional stability of double inlet systems”. *Estuarine, Coastal and Shelf Science* 107: 69–80. DOI: 10.1016/j.ecss.2012.04.033.

Brown, E.I. (1928). “Inlets on sandy coasts”. In: *Proceedings of the American Society of Civil Engineers*. Vol. 54. 1, pp. 505–553.

Bruun, P. (1981). *Port Engineering*. Houston, Texas: Gulf Publishing Company, 436 pp.

Bruun, P. and F. Gerritsen (1959). “Natural bypassing of sand at coastal inlets”. *Journal of the Waterways and Harbours Division* 85: 75–107.

— (1960). *Stability of Coastal Inlets*. North Holland Publishing Co.

Bruun, P., A.J. Mehta, and I.G. Johnsson (1978). *Stability of Tidal Inlets: Theory and Engineering*. Elsevier Scientific Publishing Co., 510 pp.

Buchwald, V.T. (1971). “The diffraction of tides by a narrow channel”. *Journal of Fluid Mechanics* 46(3): 501–511. DOI: 10.1017/S0022112071000661.

Buijsman, M.C. and H. Ridderinkhof (2007). “Long-term ferry-ADCP observations of tidal currents in the Marsdiep Inlet”. *Journal of Sea Research* 57(4): 237–256. DOI: 10.1016/j.seares.2006.11.004.

Byrne, R.J., P.A. Bullock, and D.G. Tyler (1975). “Response characteristics of a tidal inlet”. In: *Estuarine Research*. Ed. by L.E. Cronin. Vol. 2. Academic Press, pp. 267–276.

Byrne, R.J., J.T. DeAlteris, and Bullock (1974). “Channel stability in tidal inlets: a case study”. In: *Proceedings of the 14th International Conference on Coastal Engineering*. Ed. by M.P. O’Brien. Vol. 1. ASCE, pp. 1585–1604.

Byrnes, M.R., J.L. Bakker, and N.C. Kraus (2003). “Coastal sediment budgets for Grays Harbor, Washington”. In: *Coastal Sediments ’03*, pp. 1–10.

Cheung, K.F., F. Gerritsen, and J. Cleveringa (2007). “Morphodynamics and sand bypassing at Ameland inlet, the Netherlands”. *Journal of Coastal Research* 23(1): 106–118. DOI: 10.2112/04-0403.1.

Clausner, J.E., J.A. Gebert, A.T. Rambo, and K.D. Watson (1991). “Sand bypassing at Indian River Inlet, Delaware”. In: *Coastal Sediments ’91*, pp. 1177–1191.

Cleary, W.J. and D.M. FitzGerald (2003). “Tidal inlet response to natural sedimentation processes and dredging-induced tidal prism changes: Mason Inlet, North Carolina”. *Journal of Coastal Research* 19(4): 1018–1025.

Cooper, J.A.G. (2001). “Geomorphological variability among microtidal estuaries from the wave-dominated South African coast”. *Geomorphology* 40(1–2): 99–122. DOI: 10.1016/S0169-555X(01)00039-3.

Dabees, M.A., B.D. Moore, and K.K. Humiston (2011). “Evaluation of inlets channel migration and management practices in southwest Florida”. In: *Coastal Sediments ’11*, pp. 484–496.

Davis, R.A. (1994). *Geology of Holocene Barrier Island Systems*. Berlin Heidelberg: Springer, 464 pp. DOI: 10.1007/978-3-642-78360-9.

Davis, R.A. and D.M. FitzGerald (2004). *Beaches and Coasts*. Blackwell Publishing Ltd., 419 pp.

Davis, R.A. and M.O. Hayes (1984). “What is a wave-dominated coast?” *Marine Geology* 60: 313–329. DOI: 10.1016/S0070-4571(08)70152-3.

Davis, R.A., A.C. Hine, and M.J. Bland (1987). “Midnight pass, Florida: inlet instability due to man-related activities in Little Sarasota Bay”. In: *Coastal Sediments ’87*. Ed. by N.C. Kraus. ASCE, pp. 2062–2077.

DeAlteris, J.T. and R.J. Byrne (1975). “The recent history of Wachapreague Inlet, Virginia”. In: *Estuarine Research*. Ed. by L.E. Cronin. Vol. 2. New York: Academic Press, pp. 167–183.

Dean, R.G. (1988). "Sediment interaction at modified coastal inlets; processes and policies". In: *Hydrodynamics and Sediment Dynamics of Tidal Inlets*. Ed. by D.G. Aubrey and L. Weishar. Vol. 29. Lecture Notes on Coastal and Estuarine Studies. New York: Springer-Verlag, pp. 412–439. DOI: 10.1029/LN029p0412.

— (1991). "Equilibrium beach profiles: characteristics and applications". *Journal of Coastal Research* 7(1): 53–84.

Dean, R.G. and R.A. Dalrymple (2002). *Coastal Processes with Engineering Applications*. Cambridge University Press, 475 pp.

Dean, R.G. and P.A. Work (1993). "Interaction of navigational entrances with adjacent shorelines". *Journal of Coastal Research Special Issue* 18: 91–110.

Dean-Rosati, J. (2005). "Concepts in sediment budgets". *Journal of Coastal Research* 21(2): 307–322. DOI: 10.2112/02-475A.1.

de Swart, H.E. and N.D. Volp (2012). "Effects of hypsometry on the morphodynamic stability of single and multiple tidal inlet systems". *Journal of Sea Research* 74: 35–44. DOI: 10.1016/j.seares.2012.05.008.

de Swart, H.E. and J.T.F. Zimmerman (2009). "Morphodynamics of tidal inlet systems". *Annual Review of Fluid Mechanics* 41: 203–229. DOI: 10.1146/annurev.fluid.010908.165159.

Dieckmann, R., M. Osterhun, and H.W. Partenscky (1988). "A comparison between German and North American tidal inlets". In: *Proceedings of the 21st International Conference on Coastal Engineering*. Ed. by B.L. Edge. Vol. 1. ASCE, pp. 2681–2691.

Dillingh, D. (2013). *Kenmerkende waarden Kustwateren en Grote Rivieren*. Report 1207509-000-ZKS-0010. Delft, The Netherlands: Deltares.

DiLorenzo, J.L. (1988). "The overtide and filtering response of small inlet/bay systems". In: *Hydrodynamics and Sediment Dynamics of Tidal Inlets*. Ed. by D.G. Aubrey and L. Weishar. Vol. 29. Lecture Notes on Coastal and Estuarine Studies. New York: Springer New York, pp. 24–53. DOI: 10.1007/978-1-4757-4057-8_2.

Di Silvio, G. (1989). "Modelling of the morphological evolution of tidal lagoons and their equilibrium configuration". In: *Proceedings of the 13th Congress of the IAHR*, pp. 169–175.

Dombrowsky, M.R. and A.J. Mehta (1993). "Inlets and management practices: southeast coast of Florida". *Journal of Coastal Research* (SI 18): 29–57.

Dronkers, J. (1986). "Tidal asymmetry and estuarine morphology". *Netherlands Journal of Sea Research* 20(2–3): 117–131. DOI: 10.1016/0077-7579(86)90036-0.

Dronkers, J., (2005). *Dynamics of coastal systems*. World Scientific Publishing Co., 25: 519 pp.

Dronkers, J.J. (1964). *Tidal Computations in Rivers and Coastal Waters*. Amsterdam: North Holland Publishing Co., 518 pp.

— (1968). "Discussion of 'Water-level fluctuations and flow in tidal inlets' by J. van de Kreeke". *Journal of the Waterways and Harbors Division* 94(WW3): 376–377.

— (1975). "Tidal theory and computations". In: *Advances in Hydroscience*. Vol. 10. Academic Press Inc., pp. 145–230.

Ehlers, J. (1988). *The Morphodynamics of the Wadden Sea*. Rotterdam: Balkema, 397 pp.

El-Ashry, M.T and Wanless, H.R. (1965). Birth and early growth of a tidal delta. *Journal of Geology* 73: 404–406.

Escoffier, F.F. (1940). "The stability of tidal inlets". *Shore and Beach* 8(4): 111–114.

Escoffier, F.F. and T.L. Walton (1979). "Inlet stability solutions for tributary inflow". *Journal of the Waterway, Port, Coastal and Ocean Division* 105(WW4): 341–355.

Esri, DigitalGlobe, GeoEye, Earthstar Geographics, CNES/Airbus DS, USDA, USGS, AEX, Getmapping, Aerogrid, IGN, IGP, swisstopo, and the GIS User Community (2016). *ArcGIS Online*. URL: www.arcgis.com/home/webmap/viewer.html?useExisting=1.

Ferreira, Ó. (2011). "Morphodynamic impact of inlet relocation to the updrift coast: Ancão peninsula (Ría Formosa, Portugal)". In: *Coastal Sediments '11*, pp. 497–504.

FitzGerald, D.M. (1984). "Interactions between the ebb-delta and landward shoreline: Price Inlet, South Carolina". *Journal of Sedimentary Research* 54(4): 1301–1318. DOI: 10.1306/212F85C6-2B24-11D7-8648000102C1865D.

— (1988). "Shoreline erosional-depositional processes associated with tidal inlets". In: *Hydrodynamics and Sediment Dynamics of Tidal Inlets*. Ed. by D.G. Aubrey and L. Weisher. Vol. 29. Lecture Notes on Coastal and Estuarine Studies. New York: Springer-Verlag, pp. 186–225. DOI: 10.1029/LN029p0186.

— (1996). "Geomorphic variability and morphological and sedimentological controls on tidal inlets". *Journal of Coastal Research* (SI 23): 47–71. DOI: /10.2307/25736068.

FitzGerald, D.M., N.C. Kraus, and E.B. Hands (2000). *Natural Mechanisms of Sediment Bypassing at Tidal Inlets*. ERDC/CHL CHETN-IV-30. US Army Corps of Engineers, 10 pp.

FitzGerald, D.M. and D. Nummedal (1983). "Response characteristics of an ebb-dominated tidal inlet channel". *Journal of Sedimentary Research* 53(3): 833–845.

Fitzgerald, D.M., S. Penland, and D. Nummedal (1984). "Control of barrier island shape by inlet sediment bypassing: East Frisian Islands, West Germany". *Marine Geology* 60(1–4): 355–376. DOI: 10.1016/0025-3227(84)90157-9.

Friedrichs, C.T. and D.G. Aubrey (1988). "Non-linear distortion in shallow well-mixed estuaries; a syntheses". *Estuarine Coastal and Shelf Science* 27(5): 521–545. DOI: 10.1016/0272-7714(88)90082-0.

Friedrichs, C.T., D.G. Aubrey, G.S. Giese, and P.E. Speer (1993). "Hydrodynamical modeling of a multiple-inlet estuary/barrier system: insight into tidal inlet formation and stability". In: *Formation and Evolution of Multiple Tidal Inlets*. Ed. by D.G. Aubrey and G.S. Giese. Vol. 44, pp. 95–112. DOI: 10.1029/CE044p0095.

Fry, V.A. and D.G. Aubrey (1990). "Tidal velocity asymmetries and bedload transport in shallow embayments". *Estuarine, Coastal and Shelf Science* 30(5): 453–473. DOI: 10.1016/0272-7714(90)90067-2.

Gaudiano, D.J. and T.W. Kana (2001). "Shoal bypassing in mixed energy inlets: geomorphic variables and empirical predictions for nine South Carolina inlets". *Journal of Coastal Research* 17(2): 280–291. DOI: 10.2307/4300178.

Glaeser, J.D. (1978). "Global distribution of barrier islands in terms of tectonic setting". *Journal of Geology* 86(3): 283–297.

Goor, M. van (2003). "Impact of sea-level rise on the morphological equilibrium state of tidal inlets". *Marine Geology* 202(3–4): 211–227. DOI: 10.1016/S0025-3227(03)00262-7.

Hanisch, J. (1981). "Sand transport in the tidal inlet between Wangerooge and Spiekeroog (W. Germany)". In: *Holocene Marine Sedimentation in the North Sea Basin: Special Publication 5 of the IAS*. Ed. by S.-D. Nio, R.T.E. Schüttenhelm, and T.C.E. van Weering. Wiley, pp. 175–185. DOI: 10.1002/9781444303759.ch13.

Hayes, M.O. (1977). "Development of Kiawah Island, SC". In: *Coastal Sediments '77*. ASCE, pp. 828–847.

— (1979). "Barrier island morphology as a function of tidal and wave regime". In: *Barrier Islands*. Ed. by S.P. Leatherman. New York: Springer-Verlag.

— (1980). “General morphology and sediment patterns in tidal inlets”. *Sedimentary Geology* 26(1–3): 139–156. DOI: 10.1016/0037-0738(80)90009-3.

— (1994). “The Georgia Bight barrier system”. In: *Geology of Holocene Barrier Island Systems*. Ed. by R.A. Davis. Springer-Verlag, pp. 233–304.

Heath, R.A. (1975). “Stability of some New Zealand coastal inlets”. *New Zealand Journal of Marine and Freshwater Research* 9(4): 449–457. DOI: 10.1080/00288330.1975.9515580.

Herman, A. (2007). “Numerical modelling of water transport processes in partially-connected tidal basins”. *Coastal Engineering* 54(4): 297–320. DOI: 10.1016/j.coastaleng.2006.10.003.

Hicks, M.D. and T.M. Hume (1996). “Morphology and size of ebb tidal deltas at natural inlets on open-sea and pocket-beach coasts, North Island, New Zealand”. *Journal of Coastal Research* 12(1): 47–63.

— (1997). “Determining sand volumes and bathymetric change on an ebb delta”. *Journal of Coastal Research* 13(2): 407–416.

Hicks, M.D., T.M. Hume, A. Swales, and M.O. Green (1999). “Magnitudes, spacial extent, time scales and causes of shoreline change adjacent to an ebb tidal delta, Katikati Inlet, New Zealand”. *Journal of Coastal Research* 15(1): 220–240.

Hine, A.C. (1975). “Bedform distribution and migration patterns on tidal deltas in the Chatham Harbor estuary, Cape Cod, Massachusetts”. In: *Estuarine Research: Geology and engineering*. Ed. by L.E. Cronin. Vol. 2. Acad, pp. 235–252.

Hinwood, J.B. and E.J. McLean (2001). “Monitoring and modeling tidal regime changes following inlet scour”. *Journal of Coastal Research* (SI 34): 449–458.

— (2015a). “Estuaries, tidal inlets, Escoffier, O’Brien and geomorphic attractors”. In: *Proceedings of the 36th IAHR World Congress*, pp. 2344–2356.

— (2015b). “Predicting the dynamics of intermittently closed/open estuaries using attractors”. *Coastal Engineering* 99: 64–72. ISSN: 0378-3839. DOI: 10.1016/j.coastaleng.2015.02.008.

Hinwood, J.B., E.J. McLean, and B.C. Wilson (2012). “Non-linear dynamics and attractors for the entrance state of a tidal estuary”. *Coastal Engineering* 61: 20–26. ISSN: 0378-3839. DOI: 10.1016/j.coastaleng.2011.11.007.

Hume, T.M. and C.E. Herdendorf (1992). “Factors controlling tidal inlet characteristics on low drift coasts”. *Journal of Coastal Research* 8(2): 355–375.

Ippen, A.T. (1966). *Estuary and Coastline Hydrodynamics*. McGraw-Hill Book Co., pp. 505–510.

Israel, C.G. and D.W. Dunsbergen (1999). “Cyclic morphological development of the Ameland Inlet”. In: *Proceedings of the 1st IAHR Symposium on River, Coastal and Estuarine Morphodynamics (RCEM) 1999*. Vol. 2, pp. 705–714.

Jarrett, J.T. (1976). *Tidal Prism-Inlet Area Relationships*. GITI Report 3. Vicksburg, MS: U.S. Army Coastal Engineering Research Center.

Jelgersma, S. (1983). “The Bergen Inlet, transgressive and regressive Holocene shoreline deposits in the northwestern Netherlands”. *Geologie en Mijnbouw* 62(3): 471–486.

Kamphuis, J.W. (2006). *Introduction to Coastal Engineering and Management*. World Scientific Publishing Co., pp. 280–297.

Kana, T.W. (1989). “Erosion and beach restoration at Seabrook Island, South Carolina”. *Shore and Beach* 57(3): 3–18.

Kana, T.W. and J.E. Mason (1988). “Evolution of an ebb tidal delta after an inlet relocation”. In: *Hydrodynamics and Sediment Dynamics of Tidal Inlets*. Ed. by D.G. Aubrey and L. Weishar. Vol. 29. Lecture Notes on Coastal and Estuarine Studies. New York: Springer-Verlag, pp. 382–412. DOI: 10.1029/LN029p0382.

Kana, T.W. and P.A. McKee (2003). "Relocation of Captain Sams Inlet – 20 years later". In: *Coastal Sediments '03*. ASCE, pp. 1–13.

Keulegan, G.H. (1951). *Third Progress Report on Tidal Flow in Entrances: Water Level Fluctuations of Basins in Communication with Seas*. Report No. 1146. Washington, DC: National Bureau of Standards, 28 pp.

— (1967). *Tidal Flow in Entrances; Water-Level Fluctuations of Basins in Communication with Seas*. Technical Bulletin No. 14. Vicksburg, MS: U.S. Army Engineer Waterways Experiment Station, 100 pp.

King, C.A.M. (1972). *Beaches and Coasts*. Butler and Tanner Ltd., 570 pp.

Kjerfve, B. (1986). "Comparative oceanography of coastal lagoons". In: *Estuarine Variability*. Ed. by D.A. Wolfe. Academic Press, pp. 63–81.

Komar, P.D. (1998). *Beach Processes and Sedimentation*. Prentice Hall, p. 276.

Kragtwijk, N.G., T.J. Zitman, M.J.F. Stive, and Z.B. Wang (2004). "Morphological response of tidal basins to human interventions". *Coastal Engineering* 51(3): 207–221. ISSN: 0378-3839. DOI: 10.1016/j.coastaleng.2003.12.008.

Kraus, N.C. (1998). "Adaptation of the Frisian Inlet to a reduction in basin area". In: *Proceedings of the 25th International Conference on Coastal Engineering*. Ed. by B.L. Edge. Vol. 3. ASCE, pp. 3265–3278.

Kraus, N.C. (2000). "Reservoir model of ebb-tidal shoal evolution and sand bypassing". *Journal of Waterway, Port, Coastal and Ocean Engineering* 126(6): 305–313.

— (2005). "Coastal inlet functional design: anticipating morphological response". In: *Coastal Dynamics '05*. Ed. by A. Sanchez-Arcilla. ASCE, pp. 1–13. DOI: 10.1061/40855(214)108.

Kraus, N.C., L. Lin, B.K. Batten, and G.L. Brown (2006). *Matagorda Shipping Channel, Texas: Jetty Stability Study*. ERDC/CHL TR-06-7. US Army Corps of Engineers, Engineering Research and Development Center.

Kraus, N.C. and J.D. Rosati (1999). "Estimating uncertainty in coastal inlet sediment budgets". In: *Proceedings of the 12th Annual National Conference on Beach Preservation Technology*. Florida Shore and Beach Preservation Association, pp. 287–302.

Kreeke, J. van de (1967). "Water-level fluctuations and flow in tidal inlets". *Journal of the Waterways, Harbor and Coastal Engineering Division* 93(WW4): 97–106.

— (1985). "Stability of tidal inlets – Pass Cavallo, Texas". *Estuarine, Coastal and Shelf Science* 21(1): 33–43. DOI: 10.1016/0272-7714(85)90004-6.

— (1990a). "Can multiple tidal inlets be stable?" *Estuarine, Coastal and Shelf Science* 30(3): 261–273. DOI: 10.1016/0272-7714(90)90051-R.

— (1990b). "Stability analysis of a two-inlet bay system". *Coastal Engineering* 14(6): 481–497. DOI: 10.1016/0378-3839(90)90031-Q.

— (1992). "Stability of tidal inlets; Escoffier's analysis". *Shore and Beach* 60(1): 9–12.

— (1996). "Morphological changes on a decadal time scale in tidal inlets: modeling approaches". *Journal of Coastal Research* (SI 23): 73–81.

— (1998). "Adaptation of the Frisian Inlet to a reduction in basin area with special reference to the cross-sectional area". In: *Proceedings of the 8th Conference on Physics of Estuaries and Coastal Seas (PECS) 1996*. Ed. by J. Dronkers and M.B.A.M. Scheffers, pp. 355–362.

— (2004). "Equilibrium and cross-sectional stability of tidal inlets: application to the Frisian Inlet before and after basin reduction". *Coastal Engineering* 51(5–6): 337–350. DOI: 10.1016/j.coastaleng.2004.05.002.

— (2006). "An aggregate model for the adaptation of the morphology and sand bypassing after basin reduction of the Frisian Inlet". *Coastal Engineering* 53(2–3): 255–263. DOI: 10.1016/j.coastaleng.2005.10.013.

Kreeke, J. van de, R.L. Brouwer, T.J. Zitman, and H.M. Schuttelaars (2008). "The effect of a topographic high on the morphological stability of a two-inlet bay system". *Coastal Engineering* 55(4): 319–332. DOI: 10.1016/j.coastaleng.2007.11.010.

Kreeke, J. van de and D.W. Dunsbergen (2000). "Tidal asymmetry and sediment transport in the Frisian Inlet". In: *Proceedings of the 9th Conference on Physics of Estuaries and Coastal Seas (PECS) 1998*. Ed. by T. Yanagi. Tokyo: Terra Scientific Publishing Company, pp. 139–159.

Kreeke, J. van de and A. Hibma (2005). "Observations on silt and sand transport in the throat section of the Frisian Inlet". *Coastal Engineering* 52(2): 159–175. DOI: 10.1016/j.coastaleng.2004.10.002.

Kreeke, J. van de and K. Robaczewska (1993). "Tide-induced residual transport of coarse sediment; application to the Ems estuary". *Netherlands Journal of Sea Research* 31(3): 209–220. DOI: 10.1016/0077-7579(93)90022-K.

Lam, N.T. (2009). "Hydrodynamics and morphodynamics of seasonally forced tidal inlet systems". PhD thesis. Delft University of Technology, 142 pp.

LeConte, L.J. (1905). "Discussion of 'Notes on the improvement of river and harbour outlets in the United States' by D.A. Watts". *Transactions of the American Society of Civil Engineers* LV(2): 306–308.

LeProvost, C. (1991). "Generation of overtides and compound tides (review)". In: *Tidal Hydrodynamics*. Ed. by B.B. Parker, pp. 269–295.

Lesser, G.R., J.A. Roelvink, J.A.T.M. van Kester, and G.S. Stelling (2004). "Development and validation of a three-dimensional morphological model". *Coastal Engineering* 51(8–9): 883–915. ISSN: 0378-3839. DOI: http://dx.doi.org/10.1016/j.coastaleng.2004.07.014.

Lorentz, H.A. (1926). *Verslag Staatscommissie Zuiderzee 1918–1926*. In Dutch. Den Haag.

Louters, T. and F. Gerritsen (1994). *The Riddle of the Sands; A Tidal System's Answer to a Rising Sea Level*. Report RIKZ-94.040. Ministry of Transport, Public Works and Water Management, Directorate-General of Public Works and Water Managements, National Institute for Coastal and Marine Management, 69 pp.

Marino, J.N. and A.J. Mehta (1988). "Sediment trapping at Florida's east coast inlets". In: *Hydrodynamics and Sediment Dynamics of Tidal Inlets*. Ed. by D.G. Aubrey and L. Weishar. Vol. 29. Lecture Notes on Coastal and Estuarine Studies. New York: Springer-Verlag, pp. 284–296. DOI: 10.1029/LN029p0284.

Matias, A., A. Vila-Concejo, Ó Ferreira, and J.M.A. Dias (2009). "Sediment dynamics of barriers with frequent overwash". *Journal of Coastal Research* 25(3): 768–780.

McLean, E.J. and J.B. Hinwood (2000). "Modelling entrance resistance in estuaries". In: *Proceedings of the 27th International Conference on Coastal Engineering*. Ed. by B.L. Edge. Vol. 4. ASCE, pp. 3446–3457. DOI: 10.1061/40549(276)268.

Mehta, A.J. and E. Özsoy (1978). "Inlet hydraulics". In: *Stability of Tidal Inlets: Theory and Engineering*. Ed. by P. Bruun. Amsterdam, The Netherlands: Elsevier Scientific Publishing Co., pp. 83–161.

Morris, B.D., M.A. Davidson and D.A. Huntley (2004). "Estimates of the seasonal morphological evolution of the Barra Nova Inlet using video techniques". *Continental Shelf Research* 24(2): 263–278. DOI: 10.1016/j.csr.2003.09.009.

Morris, B.D. and I.L. Turner (2010). "Morphodynamics of intermittently open-closed coastal lagoon entrances: new insights and conceptual model". *Marine Geology* 271(1–2): 55–66. DOI: 10.1016/j.margeo.2010.01.009.

Murray, A.B. (2003). "Contrasting the goals, strategies, and predictions associated with simplified numerical models and detailed simulations". In: *Prediction in*

Geomorphology. Ed. by P.R. Wilcock and R.M. Iverson. Vol. 135. Geophysical Monograph Series. AGU. Washington DC, pp. 151–165. DOI: 10.1029/GM135.

Nahon, A., X. Bertin, A.B. Fortunato, and A. Oliveira (2012). "Process-based 2DH morphodynamic modeling of tidal inlets: A comparison with empirical classifications and theories". *Marine Geology* 291–294: 1–11. DOI: 10.1016/j.margeo.2011.10.001.

NASA, GSFC, MITI, ERSDAC, JAROS, and U.S./Japan ASTER Science Team (2003). *Venice, Italy*. URL: http://earthobservatory.nasa.gov/IOTD/view.php?id=3827.

National Park Service (2012). *Post-Hurricane Sandy: Old Inlet Breach on Fire Islands*. URL: www.nps.gov/fiis/naturescience/post-hurricane-sandy-breaches.htm.

O'Brien, M.P. (1931). "Estuary tidal prism related to entrance areas". *Civil Engineering* 1(8): 738–739.

— (1969). "Equilibrium flow areas of inlets on sandy coasts". *Journal of the Waterways and Harbors Division* 95(WW1): 43–52.

O'Brien, M.P. and R.G. Dean (1972). "Hydraulics and sedimentary stability of coastal inlets". In: *Proceedings of the 14th International Conference on Coastal Engineering*. Ed. by M.P. O'Brien. Vol. 2, pp. 761–780.

Pacheco, A., Ó. Ferreira, and J.J. Williams (2011). "Long-term morphological impacts of the opening of a new inlet on a multiple inlet system". *Earth Surface Processes and Landforms* 36(13): 1726–1735. DOI: 10.1002/esp.2193.

Pacheco, A., A. Vila-Concejo, Ó. Ferreira, and J.A. Dias (2007). "Present hydrodynamics of Ancão Inlet, 10 years after its relocation". In: *Coastal Sediments '07*. Ed. by N.C. Kraus and J. Dean-Rosati. ASCE, pp. 1557–1570. DOI: 10.1061/40926(239)120.

— (2008). "Assessment of tidal inlet evolution and stability using sediment budget computations and hydraulic parameter analysis". *Marine Geology* 247(1–2): 104–127. DOI: 10.1016/j.margeo.2007.07.003.

Pingree, R.D. and D.K. Griffiths (1979). "Sand transport paths around the British Isles resulting from M2 and M4-tidal interactions". *Journal of the Marine Biological Association of the United Kingdom*, 59(2): 497–513. DOI: 10.1017/S0025315400042806.

Powell, M.A., R.J. Thieke, and A.J. Mehta (2006). "Morphodynamic relationships for ebb and flood delta volumes at Florida's tidal entrances". *Ocean Dynamics* 56(3): 295–307. DOI: 10.1007/s10236-006-0064-3.

Ranasinghe, R. and C. Pattiaratchi (1999). "The seasonal closure of tidal inlets: Wilson Inlet – a case study". *Coastal Engineering* 37(1): 37–56. DOI: 10.1016/S0378-3839(99)00007-1.

— (2003). "The seasonal closure of tidal inlets: causes and effects". *Coastal Engineering Journal* 45(4): 601–627. DOI: 10.1142/S0578563403000919.

Ranasinghe, R., C. Pattiaratchi, and G. Masselink (1999). "A morphodynamic model to simulate the seasonal closure of tidal inlets". *Coastal Engineering* 37(1): 1–36. DOI: 10.1016/S0378-3839(99)00008-3.

Reid, R.O. and B.R. Bodine (1969). "Numerical model for storm surges in Galveston Bay". *Journal of the Waterways and Harbors Division* 94(WW1): 35–59.

Ridderinkhof, H. and J.T.F. Zimmerman (1992). "Chaotic stirring in a tidal system". *Science* 258(5085): 1107–1111. DOI: 10.1126/science.258.5085.1107.

Ridderinkhof, W., H.E. de Swart, M. van der Vegt, N.C. Alebregtse, and P. Hoekstra (2014). "Geometry of tidal inlet systems: a key factor for the net sediment transport in tidal inlets". *Journal of Geophysical Research: Oceans* 119(10): 6988–7006. DOI: 10.1002/2014JC010226.Received.

Rijn, L. van (1993). *Principles of Sediment Transport in Rivers, Estuaries and Coastal Seas*. Part I. Aqua Publications, 500 pp.

Roelvink, J.A. (2006). "Coastal morphodynamic evolution techniques". *Coastal Engineering* 53(2–3): 277–287. DOI: 10.1016/j.coastaleng.2005.10.015.

Roelvink, J. A. and Reniers, A. J. H. M., (2010). *A Guide to Modeling Coastal Morphology*. World Scientific Publishing Co., 12: 274 pp. DOI: 10.1142/7712.

Roelvink, J.A. and D.J.R. Walstra (2004). "Keeping it simple by using complex models". In: *Proceedings of Advances in Hydro-Science and -Engineering*. Vol. 6, pp. 1–11.

Roos, P.C. and H.M. Schuttelaars (2011). "Influence of topography on tide propagation and amplification in semi-enclosed basins". *Ocean Dynamics* 61(1): 21–38. DOI: 10.1007/s10236-010-0340-0.

Roos, P.C., H.M. Schuttelaars, and R.L. Brouwer (2013). "Observations on barrier island length explained using an exploratory morphodynamic model". *Geophysical Research Letters* 40(16): 4338–4343. DOI: 10.1002/grl.50843.

Rosati, J.D. and N.C. Kraus (1999). *Formulation of Sediment Budgets at Inlets*. Coastal Engineering Technical Note CETN IV-15. Vicksburg (MS): US Army Engineer Research, Development Center, Coastal, and Hydraulics Laboratory, 20 pp.

Salles, P., G. Voulgaris, and D.G. Aubrey (2005). "Contribution of nonlinear mechanisms in the persistence of multiple tidal inlet systems". *Estuarine, Coastal and Shelf Science* 65(3): 475–491. DOI: 10.1016/j.ecss.2005.06.018.

Seabergh, W.C. (2002). "Hydrodynamics of Tidal Inlets". In: *Coastal Engineering Manual*. 1110-2-1100. Washington, DC: US Army Corps of Engineers. Chap. II-6.

Seabergh, W.C. and N.C. Kraus (2003). "Progress in management of sediment bypassing at coastal inlets; natural bypassing, weir jetties, jetty spurs, and engineering aids in design". *Coastal Engineering Journal* 45(4): 533–563. DOI: 10.1142/S0578563403000944.

Serrano, D., E. Ramírez-Félix, and A. Valle-Levinson (2013). "Tidal hydrodynamics in a two-inlet coastal lagoon in the Gulf of California". *Continental Shelf Research* 63:1–12. DOI: 10.1016/j.csr.2013.04.038.

Sha, L.P. (1989). "Variation in ebb-delta morphologies along the West and East Frisian Islands, the Netherlands and Germany". *Marine Geology* 89(1–2): 11–28. DOI: 10.1016/0025-3227(89)90025-X.

Shigemura, T. (1980). "Tidal prism - throat area relationships of the bays of Japan". *Shore and Beach* 48(3): 30–35.

Sorensen, R.M. (1977). *Procedures for Preliminary Analysis of Tidal Inlet Hydraulics and Stability*. Coastal Engineering Technical Aid No. 77-8. CERC.

Soulsby, R. (1997). *Dynamics of Marine Sands: A Manual for Practical Applications*. Thomas Telfordt, 249 pp.

Southgate, H.N. (1993). "The effect of wave event sequencing on long-term beach response". In: *Proceedings of Large Scale Coastal Behavior '93*. Ed. by J.H. List. US Geological Survey Open-File Report 93-381. St. Petersburg, FL, 238 pp.

Spanhoff, R., E. Biegel, J. van de Graaff, and P. Hoekstra (1997). "Shoreface nourishments at Terschelling, the Netherlands: Feeder berm or breaker berm?" In: *Coastal Dynamics '97*. New York: ASCE, pp. 863–872.

Stauble, D.K. (1993). "An overview of southeast Florida inlet morphogdynamics". *Journal of Coastal Research* (SI 18): 1–27.

Stive, M.J.F., Z.B. Wang, M. Capobianco, P. Ruol, and M.C. Buijsman (1998). "Morphodynamics of a tidal lagoon and the adjacent coast". In: *Proceedings of the 8th Conference on Physics of Estuaries and Coastal Seas (PECS) 1996*. Ed. by J. Dronkers and M.B.A.M. Scheffers, pp. 355–362.

Stommel, H.M. and H.G. Farmer (1952). *On the nature of estuarine circulation*. Woods Hole Oceanographic Institution, 38 pp. DOI: 10.1575/1912/2032.

Suprijo, T. and A. Mano (2004). "Dimensionless parameters to describe topographical equilibrium of coastal inlets". In: *Proceedings of the 29th International Conference on Coastal Engineering*. Ed. by J. McKee Smith. Vol. 3, pp. 2531–2543. DOI: 10.1142/9789812701916_0204.

Tambroni, N. and G. Seminara (2006). "Are inlets responsible for the morphological degradation of Venice Lagoon?" *Journal of Geophysical Research* 111(F03013): 1–19. DOI: 10.1029/2005JF000334.

Townend, I. (2005). "An examination of empirical stability relationships for UK estuaries". *Journal of Coastal Research* 21(5): 1042–1063. DOI: 10.2112/03-0066R.1.

Tung, T.T. (2011). "Morphodynamics of seasonally closed coastal inlets at the central coast of Vietnam". PhD thesis. Delft University of Technology, 192 pp.

Tung, T.T., J. van de Kreeke, M.J.F. Stive, and D.-J.R. Walstra (2012). "Cross-sectional stability of tidal inlets: a comparison between numerical and empirical approaches". *Coastal Engineering* 60: 21–29. DOI: 10.1016/j.coastaleng.2011.08.005.

US Army Corps of Engineers (1984). *Shore Protection Manual*. Vol. II, 656 pp.

USGS and ESA (2011). *Reclaimed Lands*. URL: www.esa.int/spaceinimages/Images/2011/11/Reclaimed_lands.

van der Spek, A.J.F. and D.J. Beets (1992). "Mid-Holocene evolution of a tidal basin in the western Netherlands: a model for future changes in the northern Netherlands under conditions of accelerated sea-level rise?" *Sedimentary Geology* 80(3–4): 185–197. DOI: 10.1016/0037-0738(92)90040-X.

Veen, J. van (1936). "Onderzoekingen in de Hoofden in verband met de gesteldheid der Nederlandsche kust". PhD thesis. Leiden University.

Vennell, R. (2006). "ADCP measurements of momentum balance and dynamic topography in a constricted tidal channel". *Journal of Physical Oceanography* 36(2): 177–188. DOI: 10.1175/JPO2836.1.

Vila-Concejo, A., Ó. Ferreira, A. Matias, and J.M.A. Dias (2003). "The first two years of an inlet: sedimentary dynamics". *Continental Shelf Research* 23(14–15): 1425–1445. DOI: 10.1016/S0278-4343(03)00142-0.

Vila-Concejo, A., Ó. Ferreira, B.D. Morris, A. Matias, and J.M.A. Dias (2004). "Lessons from inlet relocation: example from southern Portugal". *Coastal Engineering* 51(10): 967–990. DOI: 10.1016/j.coastaleng.2004.07.019.

Vlasblom, W.J. (2003). *Designing Dredging Equipment*. University lecture notes. Delft University of Technology, 323 pp.

Walton, T.L. (2004a). "Escoffier curves and inlet stability". *Journal of Waterway, Port, Coastal, and Ocean Engineering* 130(1): 54–57. DOI: 10.1061/(ASCE)0733-950X(2004)130:1(54).

— (2004b). "Linear systems analysis approach to inlet-bay systems". *Ocean Engineering* 31(3–4): 513–522. DOI: 10.1016/j.oceaneng.2003.07.002.

Walton, T.L. and W.D. Adams (1976). "Capacity of inlet outer bars to store sand". In: *Proceedings of the 15th International Conference on Coastal Engineering*. Vol. 1. ASCE, pp. 1919–1937.

Walton, T.L. and F.F. Escoffier (1981). "Linearized solution to inlet equation with inertia". *Journal of the Waterway, Port, Coastal and Ocean Division, ASCE* 107(WW3): 191–195.

Wang, Z.B., H.J. de Vriend, M.J.F. Stive, and I.H. Townend (2008). "On the parameter setting of semi-empirical long-term morphological models for estuaries and tidal lagoons". In: *Proceedings of the 5th IAHR Symposium on River, Coastal and Estuarine Morphodynamics (RCEM) 2007*, Ed. by C. Dohmen-Janssen and S. Hulscher. Vol. 1. Leiden: Taylor & Francis/Balkema, pp. 103–111.

Wegen, M. van der, A. Dastgheib, and J.A. Roelvink (2010). "Morphodynamic modeling of tidal channel evolution in comparison to empirical PA relationship". *Coastal Engineering* 57(9): 827–837. DOI: http://dx.doi.org/10.1016/j.coastaleng.2010.04.003.

Welsh, J.M. and W.J. Cleary (2007). "Evolution of a relocated tidal inlet: Mason Inlet, NC". In: *Coastal Sediments '07*. Ed. by N.C. Kraus and J. Dean-Rosati. ASCE, pp. 1543–1555. DOI: 10.1061/40926(239)119.

Whitfield, A.K. (1992). "A characterization of southern African estuarine systems". *Journal of Aquatic Science* 18(1–2): 89–103. DOI: 10.1080/10183469.1992.9631327.

Williams, D.D., N.C. Kraus, and L.M. Anderson (2007). "Morphologic response to a new inlet, Packery Channel, Corpus Christi, Texas". In: *Coastal Sediments '07*. Ed. by N.C. Kraus and J. Dean-Rosati. ASCE, pp. 1529–1542. DOI: 10.1061/40926(239)118.

Winton, T.C. and A.J. Mehta (1981). "Dynamic model for closure of small inlets due to storm-induced littoral drift". In: *Proceedings of the 19th Congress of IAHR*. Vol. 2. 2. New Delhi, pp. 153–159.

Zimmerman, J.T.F. (1982). "On the Lorentz linearization of a quadratically damped forced oscillator". *Physics Letters A* 89A(3): 123–124. DOI: 10.1016/0375-9601(82)90871-4.

Zurmuhlen, F.H. (1957). "The sand transfer plant at Lake Worth Inlet". In: *Proceedings of the 6th International Conference on Coastal Engineering*. Ed. by J.W. Johnson, pp. 457–462.

Index

Printed in the United States
By Bookmasters